MORAL RESPONSIBILITY IN TWENTY-FIRST-CENTURY WARFARE

SUNY series in Ethics and the Challenges of Contemporary Warfare

---

Amy E. Eckert and Steven C. Roach, editors

MORAL RESPONSIBILITY IN TWENTY-FIRST-CENTURY WARFARE

Just War Theory and the Ethical Challenges of Autonomous Weapons Systems

Edited by

Steven C. Roach and Amy E. Eckert

Published by State University of New York Press, Albany

Printed in the United States of America

For information, contact State University of New York Press, Albany, NY
www.sunypress.edu

Library of Congress Cataloging-in-Publication Data

Names: Roach, Steven C., editor. | Eckert, Amy, editor.
Title: Moral responsibility in twenty-first-century warfare: just war theory and the ethical challenges of autonomous weapons systems / Steven C. Roach and Amy E. Eckert [editors].
Description: Albany : State University of New York, 2020. | Series: SUNY series in ethics and the challenges of contemporary warfare | Includes bibliographical references and index.
Identifiers: LCCN 2020001177 (print) | LCCN 2020001178 (ebook) | ISBN 9781438480015 (hardcover : alk. paper) | ISBN 9781438480022 (ebook)
Subjects: LCSH: Military weapons—Technological innovations—Moral and ethical aspects. | Weapons systems—United States—Technological innovations. | Artificial intelligence—Moral and ethical aspects. | Military robotics—Moral and ethical aspects.
Classification: LCC UF500 .M67 2020 (print) | LCC UF500 (ebook) | DDC 172/.42—dc23
LC record available at https://lccn.loc.gov/2020001177
LC ebook record available at https://lccn.loc.gov/2020001178

10 9 8 7 6 5 4 3 2 1

Contents

Illustrations

Figures

Tables

Acknowledgments

This book is the first to be published in the SUNY book series, "Ethics and the Challenges of Contemporary Warfare," which the editors of this volume recently established. We would like to thank Michael Rinella, the commissioning editor, for his marvelous support of this series and of this volume in particular. We are also grateful for the comments and suggestions from three external reviewers, who have helped tighten the focus of the volume and further integrate its chapters.

Introduction

Dual Moral Responsibility and the Ethical Challenges of Twenty-First-Century Warfare

STEVEN C. ROACH AND AMY E. ECKERT

The growing reliance on autonomous weapons systems (AWS) has spurred much debate about the ethics of warfare. In recent years, AWS has called attention to the rapidly changing dimension of modern warfare in which unmanned weapons will be able to detect targets and even attack satellite systems in space. For many, this highlights the need for more ethical thinking and theorizing about the moral dimensions and implications of AWS (see Gunkel 2018; Buhta et al. 2016; Sparrow 2012; Strawser 2013). The aim of this volume is to examine these dimensions and implications through various revisionist applications of just war theory to modern warfare, in particular AWS. In doing so, we seek to build on current applications of just war principles, including the rules of conduct in war (*jus in bello*) for AWS (Arkin 2010), and to address the moral risks and possibilities of extending just war principles to the artificial intelligence (AI) of AWS. The hope is that this will enable us, as Christian Brose (2019, 131–32) points out, "to focus more energy on making moral decisions about the intended outcomes of warfare."

Still, the challenge facing many ethicists of war, particularly just war theorists, is that AWS is a practical concept that appears to lie outside the human(ist) scope of the just war tradition. By critiquing the statist limitations of traditional and legalist just war theory and focusing attention on the individual human rights of actors (while also examining the legitimate status of other actors in the international realm, such as private military and security companies), revisionist just war theory has largely refrained from addressing questions regarding the legal and moral status of AWS (Gruszczak 2018, 34–35). Revisionist just war theory remains largely, if not exclusively, concerned with contesting international law and state responsibility to the extent that it uses these domains to legitimize *jus ad bellum* (right to war) and to enforce the codes of moral conduct during war (McMahan 2009; Fabre 2012). Seth Lazar (2017, 41) summarizes the revisionist challenge as follows:

> The archetypical traditionalist . . . is a nonreductivist collectivist who uses realistic cases. The archetypical revisionist is an individualist reductionist who uses cases involving meteors and mind control. Simplifying a little, we might unify the former positions under the heading of political philosophy approaches to just war theory and group the latter together as moral philosophy approach.

This revisionist challenge problematizes many of the assumptions on which the more traditional approach rests. Todd Burkhardt, for example, argues "that the issue of fighting with the right intention also requires us to understand the conditions for a just and lasting peace" (Burkhardt 2017, 1). Just war theory's narrow focus thus reflects the trouble with relying on conventional facts of human intelligence, discretion, and emotion to justify the ethics of going to war or preserving peace. Working beyond this restrictive focus requires us to critically understand the many ways that violence and advanced weapons systems marginalize persons. Much of this will in turn depend on the ethical engagement with different actors and AWS that allows us to contest and work beyond the limits of just war theory's revisionist and conventionalist applications.

This volume represents a critical engagement with this emerging gap(s) in logic regarding the moral responsibility of individuals, states, and AWS. As part of this engagement, it adopts the notion of dual moral responsibility, or the differing yet related notions of moral responsibility and legitimate authority, to analyze the changing roles and behaviors of various new actors in the global realm. Dual moral responsibility encapsulates the conflicts and contradictions driving the need for violent weapons and technologies to secure peace and to neutralize the effects of violence. At the same time, the changing conditions of warfare, including the rapidly advancing technology of AWS, are challenging the way we theorize and apply just war principles. If just war theory is to address this challenge, it must begin to confront and engage the biases and conventional logic that fix or restrict its ethical content, including the ideas and assumptions (e.g., gender and individual human agent) used to theorize about just war. The trouble, in other words, with relying on conventional biases to analyze AWS is that it exposes the very limits of these biases against the ethical potential of AWS, or the moral and legal status of AWS. We therefore need to take more seriously the moral implications of the adaptive capacity of AWS (to learn from their environment)—and whether this supports the moral autonomy subsumed under just war principles such as legitimate authority. As Robert Sparrow writes, "The use of such systems may render the attribution of responsibility for the actions of AWS to their operators problematic. . . . where such use of autonomous weapons seems to risk a responsibility gap and where this gap exists, it will not be plausible to hold that when a commander sends AWS into action, he or she is acknowledging the humanity of those the machines eventually kill" (Sparrow 2015, 108). If AWS are truly autonomous, then one could argue that they operate beyond the restraint of human operators, that is, they do not rise to the level of moral personality that would qualify them to bear responsibility for their actions (Roff 2013a). The problem is that in the absence of any assigned responsibility for the actions of AWS, it becomes increasingly problematic to deploy them.

A further complication of AWS is the erosion of the line between war and peace. The availability of low-level, low-risk force means an expanding gray area between all-out war and peaceful relations between states. Michael Walzer, for instance, devised new principles to understand this gray area, *jus ad vim*, or the ethics of using force, short of war (Brunstetter and Braun 2013). But this new set of rules also raises a very real danger of perpetuating violence or legitimizing the use of low-level force outside the theater of broader conflict (Enemark 2014). Diffusing low-level force, in other words, can erode the boundaries between conflict and peace. Because of the proliferation of actors, mapping the responsibilities of nonhuman actors becomes an even more daunting task in terms of legitimizing moral authority and autonomy. Both the just war tradition and the ethics of AWS, then, describe different and often opposing ways of analyzing the effects of violence and modern warfare (i.e., new actors with increased technological precision as well as privatized control).

The contributors address this tension by reevaluating the ethical constructs of just war theory, such as moral responsibility, accountability, autonomy, and rights, and by using various empirics of artificial intelligence (AI) to formulate the ethical applications of AWS. In doing so, they take up a range of thematic issues, including the deep morality of war, the reconfiguration of war ethics, the possible end of just war, a moral groundwork for robot rights and responsibilities, and the ethical uncertainty of advancing morality and accountability (i.e., compliance with the laws of war).

Just or Unjust Warfare

Just war tradition as we know it can be traced to the religious writings of Augustine, the fourth-century Catholic saint (Brunstetter and O'Driscoll 2017). Its emergence in the writings of Francisco de Vittoria, who opposed the Spanish conquest of the new world in the sixteenth century, stressed the importance of moral and personal virtue, particularly the restraint embodied in religious devotion and piety. However, in the early seventeenth century, the focus on virtue gave way to the application of legal principles in war, or a legalist approach to just war theory. Hugo Grotius, a seventeenth-century Dutch jurist, who was the most prominent proponent of this legalist approach, focused on secular international humanitarian law and its principles governing decision-making and conduct in war.

This legal framework was dramatically expanded in the nineteenth and twentieth centuries and codified in the 1949 Geneva Conventions, which stipulate the rules and procedures for treating noncombatants humanely during war. *Jus ad bellum* (right to war), *jus in bello* (rules of conduct), and *jus post bellum* (justice after war) constitute the basis of broad criteria for justifying the conduct of warring parties and administering justice after war, including just intention, legitimate cause—usually in the form of self-defense—the probability of success, legitimate/competent authority, and proportionality or the use of force that does not rise above the level of threat. Legitimate cause and proportionality serve as the morally objective guideposts for humanitarian military interventions, providing key normative constraints on harming/injuring civilians or noncombatants through the neutrality of the rules of war and state consent (see Lang, O'Driscoll, and Williams 2013).

As Dan Caldwell and Robert Williams (2006) argue, upholding these constraints is not sufficient to determine how peace will be upheld *jus post ad bellum*. The moral reality of war may be such that even proportionality and success of the war fail to justify authority, as we have seen with the insurgencies in Libya and Iraq. If anything, this seems to reflect the growing divide between a legalist approach, or the focus on international law and the state to justify *jus ad bellum*, and a revisionist perspective on just war theory, which, in critiquing traditionalism, emphasizes the individual in terms of justifying war through a uniform code of moral conduct (McMahan 2009). In short, the legalist tradition stresses how laws of war have evolved through international customary law that has imposed (legal) constraints and duties on states, including necessity and the right of self-defense.

In *Just and Unjust Wars*, Michael Walzer (2009) argues that the rules regulating the decision to go to war on humanitarian grounds were constituted by a long-standing moral tradition comprising the opinions of political leaders, philosophers, and scholars. Yet for Walzer, the moral reality of suffering never corresponded to the political will to redress such suffering. Rather, the virtues of the early legal tradition (in objectifying the rationale for just war) lost sight of the political reality shaping these principles and the attendant responsibility to treat such suffering as a condition for invoking these legal principles. In his view, such reality reflected how moral actions redounded to the political advantage of coercive powers and the values they sought to project overseas.[1] But while the mass killing of civilians may constitute a legitimate cause of armed intervention, it does not ensure that either a competent or trustworthy authority will enforce just war norms. In many respects, enforcing such norms through the UN Security Council—where the veto power of the permanent member states can rescind these morally compelling cases as we have seen with Syria—can dramatically raise the political stakes of promoting peace and security or of countering terrorism. Relying on international law to justify the right of states to defend themselves—and hence to go to war via the Law of Armed Conflict—assumes a difficult trade-off between state constraints and the moral failure to stop mass killings (Morkevicius 2018). This stems from the fact that states remain the primary actors or the sole source of moral responsibility, rather than the individual victims of violence, whose rights and protections have become subordinated to the moral responsibility of states.

Thus, the trouble with relying on state compliance to legitimize the responsibility to go to war is that we tend to downplay the importance of international human rights law. And yet in recent years, we have seen how gross violations of international human rights have justified recent decisions to launch humanitarian wars in Kosovo and Libya. Why should only states, then, be treated as the sole agents for justifying the right to wage war and validating the rules and moral conduct of the Law of Armed Conflict? In addition, how does the legalist focus expose the need for focusing more on human virtue to determine the nature of just cause? The idea is that understanding basic human rights protection involves more than simply the application of state rules, norms, and constraints against killing combatants and noncombatants in war: it also concerns the moral status of individuals.

Revisionist just theory adopts this idea to analyze the gap in logic between the legalist approach and an individual rights–based approach via the evolving codes of moral conduct (e.g., the Law of Armed Conflict). It therefore focuses on the changing conditions of individual moral responsibility, moral intention, and legitimate authority and argues that the Law of Armed Conflict should abide by international human rights law (see Brunstetter and Holeindre 2018; Burkhardt 2017; Eckert 2016; Gentry and Eckert 2014). Revisionist just war theory is also predicated on a cosmopolitan perspective of the individual, that is, the moral equality of individuals. As Cecile Fabre (2012, 7) argues, "All individuals, wherever they reside, have the aforementioned rights against everyone else, irrespective of residence." Global justice in this sense points to the gap between revisionists and traditionalists, in which revisionists reject the need for any authority beyond the individual, even though they do not deny the political reality of states and other communities.

One of the strategies used by revisionists is to stress the role played by compassion and empathy in reinforcing just war norms and furthering moral commitments. But hostile emotions can also fuel the impulsive actions that pressure and destabilize these norms. In addition, the conflicts between nonhostile and hostile emotions often compel us to dichotomize between compassion (or positive emotion) and hatred (negative emotion) and to rely on restraint to control the effects of hostile emotions (see Solomon 2007). Hostile emotions, it can be said, constitute the irrational elements of moral judgment. By the same token, positive emotions can exceed the ability to control the effects of humanitarian wars. In the Kosovo War, for instance, while upholding the basic human rights of

Kosovar civilians against the attack of states helped fuel the moral and humanitarian war, it also exposed the disparity between political and military objectives on the ground. In effect, the necessity of protecting the needs and rights of those suffering led to military measures, such as surgical strikes, that failed in most instances to support the political objectives of the war (i.e., forcing Slobodan Milosevic to surrender). Although revisionist theory enables us to understand the moral responsibility to launch wars on humanitarian grounds, it also constitutes an anthropocentric approach that has little to say about the emotive possibilities of robots or AWS.

Unsurprisingly, much of the extended focus of *jus in bello* to AWS has come from ethical scholars of AWS, such as Ronald Arkin and, to lesser extent, Robert Sparrow. For them, as we shall explain, AWS represents the possibility of eliminating the confusion and violent side effects arising from human emotions. Arkin, in fact, claims that the ethical autonomy of robots would effectively rule out the complicating influence of emotions (e.g., anger and fear). As such, the increasing presence of robotic weapons reinforces the need for reconciling these two ethical approaches and calls attention to how the increasing presence of AWS will require deeper ethical inquiry into the nonhuman sources of action and conduct. Thus, as Peter Singer (2009, 390) puts it, "By replacing human judgment with AI technologies, it becomes possible to limit the effects of war." Because artificial intelligence introduces precision weaponry, it raises the possibility of eliminating much of the violence caused by human error. The laws of war could in this manner tap into the advanced capacity of robots to carry out orders. For Arkin, AWS provide the opportunities for more precise compliance with the rules of lawful combat and just war criteria as stipulated in the Geneva Conventions.

In this way, Arkin stresses a dualistic notion of moral responsibility of human and nonhuman agents and restricts the emotive possibilities of the legal and moral status of robots' adaptive capacity. Yet it is this amoral efficacy of emotion, or absence of emotion, that remains in tension with a revisionist just war focus and that raises the difficult question of how best to use (an evolving notion) the notion of dual moral responsibility to work beyond the assumptions of robots' legal status, especially concerning the efforts to confront the legal gap between AWS and moral accountability. Indeed, as David Gunkel (2018) claims, insisting that robots cannot think also constitutes an event that precludes the normative possibility of robots' rights. Any strong and deepening engagement with these dualistic aspects of moral responsibility, then, needs to contend with this possibility.

The Ethical and Moral Challenges of High-Tech Warfare

With more and more military robots being programmed to fight and, in most cases, kill, the lethal use of AWS can be said to constitute an emerging reality in warfare. This also represents the practical opportunity to address some of the more vexing issues of indiscriminate violence and killing that have seemed to haunt conventional just war theorizing (i.e., the harm done to innocent civilians). For instance, if robots will be able to finally target combatants with greater precision, then it might be possible to avert the consequences and effects of the fog of war, such as the indiscriminate killing of noncombatants or the collateral damage of drone warfare (Enemark 2015; Kaag and Kreps 2014). This suggests that somehow such weapons may be reliably programmed to discriminate between combatants and noncombatants, which in turn would mean accepting and promoting the autonomous function of weapons systems. To borrow Sparrow's definition of an autonomous weapon, such a weapon is one "capable of being tasked with identifying possible targets and choosing which to attack without human oversight and that is sufficiently complex such that, even when it is functioning perfectly, there remains some uncertainty about which objects and/or persons, it will attack and why" (Sparrow 2015, 95).

One might argue, then, that there are no assurances that autonomous weapons, with highly complex sensors and satellite tracking devices that allow the weapons system to function with almost complete certainty, will be able to perform safely without human supervision. Even the fully autonomous self-driving car cannot entirely replicate the human ability to decipher the obstacles ahead. The difference between the self-driving sensors and human sensory experience reflects what some have called "split responsibility" (which unlike, dual responsibility, conveys unrelated sources/triggers of action), in which the human driver reacts and acts differently to avoid obstacles. In terms of lethal autonomous weapons systems (LAWS), the problem has produced similar concerns and has raised the political stakes for resolving the indiscriminate killing of noncombatants. As Singer (2009, 403) points out, "Robots have great difficulty interpreting context, and, at least, until they match humans in intelligence, it simply does not make sense to interpret a machine as having the equivalent of human rights of self-defense." There is also the (dark) possibility of robots using their developing intelligence to target humans, a fear that has been the subject of various science fiction films, such as the *Terminator* and *Bladerunner* series, and that only seems to project existing biases against robotic intelligence (i.e., that they are amoral agents). Despite this scenario, however, some countries remain resigned to develop autonomous weapons, even though many remain wary about their use in warfare.

Still, several powerful countries' military strategists are betting on this idea: namely, that autonomous robotic weapons will offer them an advantage militarily. In fact, the United States., Britain, and China have already begun research on the development of new lethal autonomous weapons systems, or advanced robotic weapons systems that carry their own sense detectors and are considered in this sense to be semiautonomous weapons (Topol 2016). In 2015, the United States, for instance, unveiled the design and section of the X-47B, a new pod-shaped aircraft that can be autonomously refueled in midair, while Britain, not to be outdone, is working on the Tauris aircraft equipped with automatic laser sensors. With nearly USD 72 billion invested in such technology, the United States continues to maintain that such advanced technology poses few risks to civilians and allows the US to better protect itself from outside threats.

The fear, then, is that AI technologies may lead to increased (human) collateral damage owing to software malfunction or programming errors. Signs of this threat have surfaced in earlier incidents involving the limited supervision of LAWS, or semiautonomous LAWS. In 2007, a South African semiautonomous antiaircraft system accidently fired upon and killed seven South African soldiers; and in 1988, the US air defense system mistakenly shot down an Iranian passenger jet (Gubrud 2016). The thorny issue is whether killer robots can be held to account for their actions; for with no human at the helm, it becomes increasingly unclear as to how to prosecute the destructive actions of robots. The only real option may be to file civil charges, effectively holding the civil programmers of these robots liable for damages. But this is not likely to curb the destructive actions of what some have called killer robots, since it will involve a high burden of proof, or depend on whether the maker had knowledge of a programming defect.

The problem with bridging this gap between international criminal law and LAWS is that AI adaptation currently remains undeveloped in relation to the rules and procedures for assigning moral responsibility. This leaves open the question of whether such responsibility can be assigned when the targeted individual lacks any sense of guilt or conscious intent (see Marchant et al. 2011). Will there be different and fair standards established for human and nonhumans, for instance? And will moral punishment for war crimes remain grounded in command responsibility (of the individual programing the robots) and, as such, displace the guilt and intent of a growing population of killer robots within the corpus of international criminal law? If there are two independently evolving tracks of human and nonhuman warfare, will the latter require a whole new conception of rights and autonomy to generate the efficacy of international criminal norms and the many new rules of procedure for determining the intent and knowledge of war criminals? The International Criminal Court and International Criminal Tribunals, for example, have brought hundreds of war criminals to justice and arguably helped deter criminal behavior (Roach 2006, 2013). But such a deterrent effect, which relies on the capacity of courts to expose the knowledge of perpetrators' intent, cannot, at least for now, apply to autonomous robots programmed to kill.

Nonetheless, many lethal weapons systems such as the PHALANX Close-in System, which is a rapid-fire computer-controlled radar gun invented by General Dynamics, are only minimally capable of firing on their own or without human guidance. Military officials are still dubious about the need to remove these and more complex semiautonomous weapons (Tauris aircraft) from human decision-making. This seems to ensure that we will continue to struggle to define the nascent, open-ended parameters of dual moral responsibility. Moreover, it might also explain the polarizing responses LAWS.

For the most part, the response to this legal and moral challenge of AI has been twofold: (1) either reject or resist LAWS or (2) devise whole new ways of rethinking their evolving role (and agency) in warfare (O'Connell 2013). The former is highlighted by concerted calls for a complete ban on LAWS by Human Rights Watch (HRW) and the Campaign to Stop Killer Robots, a coalition of nongovernmental organizations (NGOs) working to ban fully autonomous weapons (which has been in the forefront of this movement to ban all LAWS) (Campaign to Stop Killer Robots 2018). In a report issued in April 2015, HRW documented the rapid rise of many semiautonomous weapons, arguing that regulation will do little to stop the destructive impact of fully autonomous killer robots (Human Rights Watch 2015). HRW lawyers and activists recently voiced their concerns at a delegate meeting of the Convention on Certain Conventional Weapons, an agreement signed by 125 countries that has pledged to eliminate weapons that indiscriminately kill civilians (UNOG 1980).

Despite these good intentions, such resistance has done little to alter the political reality of LAWS or the most powerful countries' commitment to produce more sophisticated LAWS. As both Peter Singer and August Cole (2016) argue, it is perhaps more realistic to transition to or erect new laws and rules to hold humans accountable for any lethal mistake made by the robots. By clarifying which maker is and is not responsible, the hope is that authorities will adopt rules constraining the reckless behavior of states and corporations (Singer 2009, 20). This may be the first uncomfortable step of working toward developing and engaging the normative possibilities of their agency (or autonomy) and bridging the gap between the just war norms of legitimizing moral conduct (or just intent) and AWS. It may also be why many have sought to address the complexity of this challenge through disciplinary approaches or a multidisciplinary approach that can map an evolutionary pathway of robotic intelligence in which intention, conscience, and even emotion (feelings) might justify a new conception of (moral) responsibility and new rights (Tripod and Wolfendale 2012; Kahn 2002; Benjamin 2013; Sparrow 2012).

In short, new ethical guidelines will be needed to regulate the moral conduct of robots and to bridge the ethical gap between the Laws of Warfare and robot intelligence. Moreover, there needs to be a larger effort to work beyond existing conventions and customs of war as well as understanding the critical ties between just war theorizing and the moral autonomy of robots.

Overview of the Book

The first part of the volume focuses on the limits and problematic aspects of just war theory and the attempts to revise and contest these limits. The most compelling set of challenges to conventional just war theory involves a "deep morality of war" approach to the ethics of killing in war, which draws from cosmopolitan political theory to question the justice of a state-based system and status-quo rules of war. The contributors to this first part of the volume seek, in creative ways, to refigure the norms of moral responsibility of just war theory.

In chapter 1, Peter Sutch addresses several developments in international law and in military technology and practice that have sparked large-scale criticisms of traditional just war theory. He provides a novel defense of legalist or conventionalist just war theory against the attacks by cosmopolitan critics. Such criticisms generate *jus ad bellum* arguments for an expanded right to war for humanitarian and defensive purposes and *jus in bello* arguments that deny the moral equality of combatants that underpins the distinction between combatant and civilian. These arguments challenge the basic principles of just war theory and the laws of war in the name of a deeper and more refined moral philosophy. They also create deep practical challenges for the humane governance of conflict. The chapter focuses on two rather different modern developments—the increasing normative importance of human rights and the evolution of military technology that enables "riskless" combat—to show the differences between legalist just war theory and that of its critics. It argues that a conventionalist understanding of human rights is both more relevant to modern warfare and still an effective critical tool of managing conflict.

The detachment from the reality of violence, then, is symptomatic of the dichotomous thinking on war and peace and violence and gender. In chapter 2, Laura Sjoberg analyzes what she calls the trichotomy: namely, unjust war (the ultimate evil); just war (necessary evil and morally permissible); and just peace (that toward which just war strives). The chapter examines various assumptions of this trichotomy, including bad violence, acceptable/good violence, and nonviolence, and argues that war and peace are conceptually and empirically problematic. Building on feminist theorizing about the links between sexism, patriarchy, and violence, Sjoberg proposes both that violence is a continuum rather than a delineable entity, and that there is no nonviolent alternative to violence. After laying out its theoretical approach to violence, she turns to exploring that interpretation's implications for just war theorizing. Here, she contends that no additive or multiplicative approach to *jus ad bellum*, *jus in bello*, or *jus post bellum* can account for thinking about violence as a continuum. A continuum approach to violence, she concludes, has a number of important implications for many just war concepts, as well as for the overall utility of just war thinking.

In chapter 3, Thomas Doyle takes up this nonconventional logic of just warfare by addressing the issue of whether states' reliance on automated weapons systems in contemporary warfare motivates a similar line of inquiry into the moral responsibility related to contemporary nuclear deterrence. In his chapter, he focuses on the extent to which state practices of nuclear deterrence and their plans for responding to nuclear deterrence failure have generally entailed a sufficient loss of human autonomy by state leaders. To address this question, he undertakes a brief conceptual analysis of "autonomy" and the conditions under which it might be ceded by reliance on diverse modes of instrumental rationality. He then applies this analysis to the logic of the practice of nuclear deterrence, noting the strategic and moral problems that emerge with respect to the human control of nuclear force. Here he concludes that nuclear deterrence counts as a borderline case of "dual moral responsibility" insofar as it seems to compel at some level ethical detachment (but not in the sense of ethical neutrality) and a partial delegation of moral decision-making to "automatic" processes *while at the same time* retaining some measure of human control over the prospective uses of nuclear weapons. Under conditions of deterrence failure, the logic of nuclear reprisal suggests moral nihilism, and this, too, is in need of further theorization regarding moral responsibility in the case that nuclear weapons truly do, as Michael Walzer famously put it, "explode just war theory."

Another way of contesting the traditional or conventionalist logic of just war theory is through the legitimate role of unconventional actors that are challenging legalist just war theory. In recent years, private military and security companies have adopted strategies of seeking legitimacy under international law through the antimercenary norm. What this suggests is that just war criteria such as competent authority are being reconstituted by high-tech modern warfare. In chapter 4, Sommer Mitchell shows how the antimercenary norm has prohibited private actors from participating in conflict. Private military and security companies (PMSCs) have been hired regularly by state actors as well as international organizations to provide support services for military and security operations. This need to outsource warfare to PMSCs, she shows, reveals a complex struggle to acquire legitimacy through compliance with international, antimercenary norms. PMSCs, and not simply states, have begun to shape and reconstitute the (dispersed) meaning of legitimate authority in twenty-first-century warfare by altering perceptions of their commitment to human rights protections. The issue this raises is whether new human agents can help bridge the gap between conventional just war theory and robotic intelligence, and if this gap symptomizes the misguided faith in human fallibility of revisionist just war theorizing.

Part II consists of essays that seek to varying degrees to build on research supporting the moral responsibility of AWS or robots. In chapter 5, David Gunkel addresses the issue of whether the growing recognition of the legality of robots should make way for stronger engagement with robot rights. He makes a case for this apparently marginal set of concerns by responding to a seemingly simple and direct question: Can or should killer robots have rights? This question is not just any question. Indeed, we should be clear about the inherent difficulty of even articulating such a query, since the very concept of robots having rights strains common sense or good scientific reasoning. That it needs to be purposefully avoided as something that must not be thought, insofar as it is a kind of prohibited idea or blasphemy that would open a Pandora's box of problems and therefore should be suppressed or repressed. Gunkel argues that the existing classification schema—one that recognizes only two kinds of entities, personnel or equipment—may be too restrictive and insensitive to respond to and take responsibility for the different kinds of things with which we interact and involve ourselves. Whatever the reason(s), there is something of a deliberate decision and concerted effort not to think—or at least not to take as a serious matter for thinking—the question of robot rights. In this, Gunkel explores various reasons for considering the rights of robots in general and the rights of battlefield robots.

In chapter 6, Jai Galliott analyzes the problem of how to assign moral responsibility when large groups of people, organized or unorganized, wrongfully cause some harm that is pervasive in our world given the ubiquitous nature of large organizations, such as corporations, nations, and universities, that are involved in the development and deployment of emerging military technologies. He argues that advocates of a ban on lethal autonomous systems have erroneously attempted to take the problem of many hands one step further in suggesting that said weapons systems have or will lead us to a problem of "no hands." Here he deconstructs such arguments and, in response, recharacterizes the matter confronting lethal autonomous systems as a traditional problem of many hands when traditionally conceived as the occurrence of a situation in which the collective can reasonably be held morally responsible for an outcome, even though none of the individuals can reasonably be held morally responsible for that outcome. His aim, then, is to develop a conceptual framework for moral responsibility in cases where the problem of many hands arises in the context of the design, development, or deployment of lethal autonomous weapons that facilitates the formulation and implementation of solutions.

Chapter 7 is an attempt to test and empirically support the instrumental and constitutive features of artificial intelligence as it relates to the ethics of lethal autonomous weapons. It contends that military operations should be immune from the progress of automation and artificial intelligence evident in other areas of society. Luminary figures in science and industry as well as organized protest groups have called for an international ban on "killer robots" and the "weaponization of AI." A foundation of the argument is that to comply with international humanitarian law, autonomous weapons would need "human qualities," which, the authors argue, machines inherently lack. In contrast, the development and deployment of AI in weapons is an *ethical imperative*. A simple illustration is a weapon capable of recognizing the unexpected presence of the international protection symbol of the Red Cross in a defined target area and aborting an otherwise unrestrained human-ordered strike. This is an example of what can be called an "ethical weapon," which need not possess every human-like quality to produce a useful ethical enhancement. Ethical weapons technology is proposed to be fully integrated into the military enterprise with human commanders. The chapter outlines a case for ethical weapons and proposes a code adapted from the German Ministry of Transport and the Digital Infrastructure Ethics Commission's code for driverless automobiles. Aspects of the technological feasibility of realizing this ethical code is considered in terms of human and weapon competency, authority, and responsibility. In providing the formal semantic definitions of these concepts for encoding into weapons to provide the "meaningful human control" hitherto claimed as lacking by advocates of a ban, they seek to demonstrate the feasibility of implementing limited "human qualities" in a weapon, as part of an enterprise spanning humans and machines, in order to improve ethical outcomes.

Note

1. Thomas Hobbes's idea was that coercion was necessary to validate the contract(s) among warring parties in a state of nature.

References

Arkin, Ronald C. 2010. "The Case for Ethical Autonomy in Unmanned Systems." *Journal of Military Ethics* 9(4): 332–41.

Benjamin, Medea. 2013. *Drone Warfare: Killing by Remote Control*. London: Verso, 2013.

Brose, Christian. 2019. "The New Revolution in Military Affairs." *Foreign Affairs* 98(3): 122–34.

Brunstetter, Daniel, and Megan Braun. 2013. "From Jus ad Bellum to Jus ad Vim: Recalibrating Our Understanding of the Moral Use of Force." *Ethics & International Affairs* 27(1): 87–106. doi:10.1017/S0892679412000792.

Brunstetter, Daniel R., and John-Vincent Holeindre. 2018. The Ethics of Peace and War Revisited. Washington, DC: Georgetown University Press.

Brunstetter, Daniel R., and Cian O'Driscoll. 2017. *Just War Thinkers: From Cicero to the 21st Century*. London: Taylor and Francis.

Buhta, Nehal, Susanne Beck, Robin Geiß, Hin-Lan Yiu, and Claus Kreß. 2016. *Autonomous Weapons Systems: Law, Ethics, Policy*. Cambridge: Cambridge University Press.

Burkhardt, Todd. 2017. *Just War and Human Rights: Fighting with Right Intention*. Albany: State University of New York Press.

Caldwell, Dan, and Robert E. Williams. 2006. "Jus Post Bellum, Just War Theory and the Principles of Just Peace." *International Studies Perspectives* 7(1): 309–20.

Campaign to Stop Killer Robots. 2018. "Convergence on Retaining Control on Weapons Systems." https://www.stopkillerrobots.org/.

Curtin, Audrey Kurth. 2013. "Why Drones Fail." *Foreign Affairs* 95(5).

Eckert, Amy E. 2016. *Outsourcing War: The Just War Tradition in Age of Military Privatization*. Ithaca, NY: Cornell University Press.

Enemark, Christian. 2014. "Drones, Risk, and Perpetual Force." *Ethics & International Affairs* 28(3): 365–81.

Enemark, Christian. 2015. *Armed Drones and the Ethics of War*. New York: Routledge.

Fabre, Cecile. 2012. *Cosmopolitan War*. Oxford: Oxford University Press.

Gentry, Caron E., and Amy E. Eckert. 2014. *The Future of Just War: New Critical Essays*. Athens: University of Georgia Press.

Gruszczak, Artur. 2018. "Violence Reconsidered: Towards Postmodern Warfare." In *Technology, Ethics, and the Protocols of Modern Warfare*, edited by Artur Gruszczak and Pawel Frankowski, 26–40. New York: Routledge.

Gubrud, Mike. 2016. "Why Should We Ban Autonomous Weapons? To Survive." *IEEE Spectrum*, June 1. https://spectrum.ieee.org/automaton/robotics/military-robots/why-should-we-ban-autonomous-weapons-to-survive.

Gunkel, David. 2018. *Robots' Rights*. Cambridge, MA: MIT Press.

Human Rights Watch. 2015. "Mind the Gap: The Lack of Accountability for Killer Robots." April 9. https://www.hrw.org/report/2015/04/09/mind-gap/lack-accountability-killer-robots.

Kaag, John, and Sarah Kreps. 2014. *Drone Warfare*. London: Polity.

Kahn, Paul. 2002. "The Paradox of Riskless War." *Philosophy and Public Policy Quarterly* 22(3): 2–8.

Lang, Anthony F., Cian O'Driscoll, and John Williams. 2013. *Just War: Authority, Tradition, and Practice*. Washington DC: Georgetown University Press.

Lazar, Seth. 2017. "Just War Theory: Traditionalists versus Revisionists." *Annual Review of Political Science* 20: 37–54.

Lucas, George J. 2013. "Jus in Silico: Moral Restrictions on the Use of Cyber Warfare." In *Routledge Handbook of Ethics and War: Just War Theory in the 21st Century*, edited by Fritz Allhoff, Nicholas G. Evans, and Adam Henseke. London: Routledge.

Marchant, Gary E., Braden Allenby, Ronald Arkin, Edward T. Barret, Jason Borenstein, Lyn M. Gaudet, Orde Kittrie, Patrick Lin, George R. Lucas, Richard O' Meara, and Jared Silberman. 2011. "Institutional Governance of Autonomous Military Robots." 12 *Columbia Journal of Science and Technology Law Review* 272.

McMahan, Jeff. 2009. *Killing in War*. Oxford: Oxford University Press.

Morkevicius, Valerie. 2018. *Realist Ethics: Just War Traditions as Power Politics*. Cambridge: Cambridge University Press.

O'Connell, Mary Ellen. 2013. "Banning Autonomous Weapons." In *The American Way of Bombing*, edited by Matthew Evangelista and Henry Shue. Ithaca, NY: Cornell University Press.

Roach, Steven C. 2006. *Politicizing the International Criminal Court: The Convergence of Politics, Ethics, and Law*. Lanham, MD: Rowman & Littlefield.

Roach, Steven C. 2013. "How Political Is the ICC? Pressing Challenges and the Need for Diplomatic Efficacy." *Global Governance* 19(4): 507–23.

Roff, Heather M. 2013a. "The Strategic Robot Problem: Lethal Autonomous Weapons in War." *Journal of Military Ethics* 13(3): 211–27.

Roff, H. M. 2013b. "Responsibility, Liability, and Lethal Autonomous Robots." In *Routledge Handbook of Ethics and War: Just War Theory in the 21st Century*, edited by Fritz Allhoff, Nicholas G. Evans, and Adam Henseke, 352–64. London: Routledge.

Singer, P. W. 2009. *Wired for War: The Robotics Revolution and Conflict in the 21st Century*. New York: Penguin Books.

Singer, P. W., and August Cole. 2016. "Humans Can't Control Killer Robots, but Humans Can Be Held Accountable for Killer Robots." *ViceNews*, April 15. https://www.vice.com/en_us/article/zm75ae/killer-robots-autonomous-weapons-systems-and-accountability.

Solomon, Robert C. 2007. *True to Our Feelings: What Our Emotions Are Really Telling Us*. Oxford: Oxford University Press.

Sparrow, Robert. 2012. "War without Virtue." In *Killing by Remote Control: The Ethics of an Unmanned Military*, edited by Bradley Jay Strawser. Oxford: Oxford University Press.

Sparrow, Robert. 2015. "Robots and Respect: Assessing the Case against Autonomous Weapons Systems." *Ethics and International Affairs* 30(1): 93–116.

Strawser, Bradley Jay, ed. 2013. *Killing by Remote Control: The Ethics of an Unmanned Military*. Oxford: Oxford University Press.

Topol, Sarah. 2016. "Attack of the Killer Robots." *Buzzfeed*, August 26. https://www.buzzfeed.com/sarahatopol/how-to-save-mankind-from-the-new-breed-of-killer-robots?utm_term=.nhmMmknd1#.hnkBKe6oN.

Tripod, Paulo, and Jessica Wolfendale, eds. 2012. *New Wars and New Soldiers: Military Ethics in the Contemporary World*. London: Routledge.

United Nations Office in Geneva (UNOG). 1980. "Convention on Prohibitions or Restrictions on the Use of Certain Conventional Weapons." Geneva, October 10. https://treaties.un.org/pages/ViewDetails.aspx?src=TREATY&mtdsg_no=XXVI-2&chapter=26&clang=_en.

Walzer, Michael. 2009. *Just and Unjust Wars*. 4th ed. New York: Basic Books.

Zenko, Micha. 2016. "Do Not Believe the US Government's Official Numbers on Drone Strike Civilian Casualties." *Foreign Policy*, July 5. http://foreignpolicy.com/2016/07/05/do-not-believe-the-u-s-governments-official-numbers-on-drone-strike-civilian-casualties/.

PART I

JUST WAR AND MORAL AUTHORITY

Chapter 1

Defending Conventionalist Just War Theory in the Face of Twenty-First-Century Warfare

PETER SUTCH

This chapter mounts a critical defense of legalist just war theory. Such a defense is needed as more and more voices demand that the laws of war be radically overhauled to better reflect, so it is argued, the changing nature of war and meet the moral imperatives arising from that change. The laws of war represent one of the most important developments in what is often termed the "legalization" of world politics. The construction of global multilateral institutions like the United Nations and the creation of legal covenants with global impact have been central to the priority goal of the international community in the twentieth and twenty-first centuries. Saving humanity from the "scourge of war" has seen the prohibition on the use of aggressive force, a customary approach to the right of self-defense with restricted preemptive uses, a compulsory international mechanism for authorization of the legitimate use of force, strict rules discriminating between combatants and noncombatants (including the treatment of military personnel who when wounded or captured are hors de combat), and prohibitions on the use of weapons that cause unnecessary suffering. In each of these areas the laws of war are being challenged.

Getting to the root of this challenges requires that we confront moral and political claims about the changing nature of war. In the first part of this chapter I will argue that these challenges are grounded in a belief that contemporary conflicts, especially those associated with the "war on terror" and those fought for humanitarian purposes, have a distinct moral and thus practical character and are therefore not the same as war as provided for in the traditional laws of war. It is these beliefs that ground arguments about the need for an expanded right to self-defense (or preventative war) or for a unilateral right to humanitarian intervention. It is these arguments that can often be found at the root of policies and practices (some intended to be hidden from public view but nevertheless sanctioned at the highest levels of government) that advocate the use of torture, rendition, targeted killing using drone technology, and denial of due process. However, and this is the principal burden of this chapter, a conventionalist approach to just war is much more than a body of restrictions. It is deeply embedded in a conception of the legal and political nature of international justice and the nature of war as constituted by legalized and globalized ethics. It is neither unduly conservative nor immune to normative reform, and its restrictionist character should be embraced.

In order to really explore the claim that contemporary war is different and thus should be subject to different moral and legal rules I briefly explore two critiques of legalist just war rules. The claim that our moral response to war should be different is not only conditioned by new asymmetries and technologies. War, law, and ethics are co-constituted—with claims about the nature of ethics and morality as central to claims concerning the nature of contemporary warfare. In turning to debates on international political theory (IPT) I explore two arguments that confront the conventionalist position. The first is a revisionist theory that has become well known for its "deep morality of war" approach to the ethics of killing in war that is highly critical of some of the settled norms of the law of war (Rodin 2002; Rodin and Shue 2008; May 2008; McMahan 2008, 2009). The second is a version of cosmopolitan political theory that uses a conception of universal and individual human rights to question the justice of the existing international legal order and the established rules of war (Beitz 2009; see also Buchanan 2004, 2010, 2013; Buchanan and Keohane 2004, 2015). While these two critiques of conventionalist just war are very distinct, what makes them relevantly similar for the purposes of this chapter is the claim that legalist just war theory does not have the resources to respond to the challenges of contemporary warfare. The revisionist position has received a lot of attention in recent years and my only addition to that literature is an argument that challenges McMahan's claim that revisionist just war theory and Michael Walzer's conventionalism begin from the same moral premise. In showing that the moral projects are not comparable (and reserving an argument in favor of the conventionalist position for later sections), I turn to the cosmopolitan institutionalism of Allen Buchanan and Robert Keohane, who are, I argue, engaged in a relevantly similar project to legalist just war theory. Here I explore their basic premise and two arguments for reform to *ad bellum* and *in bello* norms and institutions. Once again I challenge the claim that legalism is unduly and structurally conservative and show how legalized moral reasoning copes with normative innovation. Finally, in showing how a conventionalist account of individual protections in international humanitarian and human rights law more adequately responds to the challenges of twenty-first-century warfare the chapter defends the ethical and legal principles of conventionalist just war theory.

The Changing Nature of War: A Brief Overview

The claim that contemporary warfare is different from that provided for in international law affects both *ad bellum* and *in bello* norms in a number of ways, many of which have direct policy implications and have underpinned changing practices in twenty-first-century warfare. The general pattern of argument is that the *ad bellum* justification for action is morally urgent, providing for expanded and often unilateral grounds for the use of force as well as a loosening of *in bello* restrictions. Two major examples of this trend are humanitarian intervention and the "war on terror." In the case of humanitarian intervention, the debates around the turn of the century focused on the obligation of the international community to respond, with force where necessary, to gross violations of human rights described as crimes against humanity. Both state practice and the jurisprudence of international criminal tribunals extended the competence of the United Nations Security Council (UNSC) to noninternational armed conflict and, in the face of UNSC deadlock, a NATO-led intervention in Kosovo was described (in the final report of the Independent International Commission on Kosovo) as illegal yet legitimate. The idea that legitimacy (grounded in the humanitarian objective of war) can be at odds with the law underpins a central debate in just war theory and practice—whether justice stems from following moral principle or law. It has driven more applied debate on the subjects of UNSC reform, negotiation on the establishment of a "Responsibility to Protect," and, in the absence of these reforms to existing law, arguments legitimizing unilateralism in breach of United Nations Charter obligations. The war on terror also moved the debate about just cause beyond established legal parameters. In the rhetoric of the president of the United States the war was necessary to confront evil. The "axis of evil" language of the 2002 National Security Strategy underpins the Bush Doctrine, which pushes the boundaries of preemptive self-defense much further than the customary Caroline/Webster formula in that it rests on the right to respond to a "grave and gathering danger," a yet unmanifested threat, and thus on a claim to preventative rather than preemptive self-defense (Doyle 2008). Not only is the customary formulation much tighter (demanding an "instant and overwhelming danger"), the Bush Doctrine sits very uneasily with just war criteria for judging the justice of the use of force, making judgments about last resort, proportionality, and reasonable chance of success more challenging (Egede and Sutch 2013, 284). It also reinstates the distinction between civilized states (those within the ambit of the law) and rogue states (Brunée and Toope 2004, 406). Not only does this reintroduce a deeply problematic division into the international legal order—privileging "rights respecting," democratic, liberal, or merely "with us" rather than "against us" states against the rest of the world—it also underpins arguments for a weakening of or the nonapplication of some *in bello* norms denying crucial humanitarian protections to both combatants and noncombatants.

The moral rightness of the *ad bellum* claim and the urgency of defeating an evil enemy has led states to revisit the peremptory norm prohibiting torture or rendition—explicitly in the case of the Landau Commission and the subsequent model endorsed by the Israeli High Court of Justice and more clandestinely in the policies adopted by the United States and the United Kingdom and their allies (Ginbar 2008; see also the Report of the Eminent Jurists Panel on Terrorism, Counter-Terrorism and Human Rights 2009). In 2002 to the end of 2003 memos from the US Department of Justice and from military lawyers maintained that enemy prisoners were considered outside of the Geneva Conventions—a challenge to the equality of combatants principle at the heart of *in bello* principles (see, for example, Sands 2008). New technologies, the clearest example of which is weaponized drones, combine the attractiveness of low risk but effective kinetic capability with the moral certainty of a just cause. They lead states more readily to violate sovereignty by engaging enemies in the territory of nonconsenting and nonbelligerent states, to overuse the military option, and they may lead to violations of the principle of discrimination as enemy fighters with a "continuous combat function" are pursued in civilian contexts (Buchanan and Keohane 2015, 18–19).

It is clear then that several key pillars of the laws of war are being directly challenged. Rather than pursue the empirical and legal questions concerning what laws have been violated we should instead turn to the foundational question. This concerns the claim that contemporary wars are different and that morality requires us to amend or violate the law in order to act justly. Where conventionalist just war theory refuses the moral claim or where the law is resistant to reform it is to be condemned.

Just War Theory: Subordinating Morality to Law

While there are some important differences between the revisionism of thinkers such as Jeff McMahan and the cosmopolitanism of scholars such as Allen Buchanan, they share a key attribute that pushes them to challenge the received wisdom of just war thinking. Central to their critique of just war theory (and the existing laws of war) is their challenge to what they see as the conservative and morally deficient understanding of the relationship between ethics and law in just war theory. Just war theorists, it is claimed, subordinate morality to the existing laws of war in ways that tie their prescriptions to outdated and immoral practices. McMahan, for example, argues that the laws of war are (and must be) substantially divergent from the morality of war (McMahan 2008, 19). He is fiercely critical of the misguided belief by contemporary just war theorists that their orthodox position "is not the result of moral theory modelling itself after the law, or being conscripted to lend its support to the authority of the law, but is instead the happy convergence of law with morality" (McMahan 2008, 20). A properly critical approach to war requires, he maintains, the participant or observer to be aware of this divergence. The limitations of the laws of war are the limitations of international law in general. Buchanan and McMahan describe such limitations as a consequence of the institutional underdevelopment of the international legal order and argue that the moral shortcomings of the laws of war are a consequence of the pragmatic conservatism of an international system geared up to prevent the hegemonic practices of the powerful rather than to promote the core values of morality (Buchanan

2010, 86; McMahan 2008, 34). This distinction between the morality of war and the laws of war is the basic element of their work that underpins a critique of a range of core principles, and both argue in favor of some morally motivated violations of the laws of war.

Both McMahan and Buchanan target Michael Walzer's classic contribution to the genre. Walzer famously redefined the traditional principles of *jus ad bellum* and *jus in bello* as the Legalist Paradigm and the War Convention (Walzer 1977). Walzer's target was the realist tradition in International Relations scholarship that argues that morality and law had no place in the science of global politics. In showing how important moral and legal judgments were to questions of war Walzer does present a theory that takes the state-centric nature of the international legal order as given and as a constraint on moral reasoning about war—indeed, about global justice more broadly. It is this that McMahan and Buchanan, albeit in different ways, find morally offensive.

Nevertheless, despite finding similar offense the two critiques are distinct in that the revisionist critique sees law as only instrumentally necessary and as immoral and the cosmopolitan-institutionalist critique sees law as deficient by its own moral standards. While much of what follows will focus on the theoretical contest between the legalist tradition and its critics, this chapter is motivated by the concern that the practical consequence of the revisionist and cosmopolitan positions is that they offer too much license to warring parties. The moral logic of these positions establishes moral hierarchies in ways that license political and legal hierarchies—between just and unjust combatants (goodies and baddies), rights-respecting and rights-violating states, civilized and noncivilized peoples—that do not seem to comprehend the legal (as opposed to the moral) complexity of the idea of responsibility (see, for example, Waldron 2018). This is not because either scholar intends to do so—indeed, McMahan's work restricts the moral right to kill so fully that his position can be described as a contingent pacifism (McMahan 2010, 44–68). Nevertheless, conceiving of the relationship between morality and law as these critiques do offers significant room for those with different, and just as closely held, metaethical doctrines, to make judgments that would cloud even the fog of war. Doctrines of humanitarian intervention, preventative self-defense, forcible democratization, the use of high-tech weaponry to pursue terrorists, or the idea that there are just and unjust combatants and that some civilians can be more justly liable to attack than some military personnel make sense in their own terms (that is, in the context of a liberal theory of individual autonomy rights). But they make little sense in the context of the politics of international law and war. Indeed, they offer significant challenges for the regulation of war and the project of mitigating its horrors and create new license for assertive liberals, tyrannical governments, and jihadist fighters.

Buchanan and McMahan both argue that if we have new, developed, global institutions (a world court of human rights, a reformed Security Council, a league of rights-respecting states, a drone accountability regime) it would be possible to bring the law of war closer to the demands of morality. Indeed, Buchanan accuses just war theory of institutional blindness. But such license, in the absence of reforms that have been so challenging in practice, is a major cause for concern. More importantly, Walzer, and others in this tradition, argue that the sorts of institutions that might render just war theory and the existing laws of war obsolete are not only too far away to contemplate, they may be morally undesirable, sacrificing legitimate moral and social pluralism for a globalized "tyranny from the center" (Walzer 2004, 188; see also Reus-Smit 2005). The rules of war are linked to a broader conception of international justice, one where human rights, peace, sovereignty, and other settled norms (Frost 1996) interact in a legalized social framework. In other words, the status quo reflects a dynamic, institutionalized conception of justice, and it is a deep understanding of this that grounds the legalist tradition.

Mapping the Debate

If we imagine a spectrum of positions in this debate the poles are taken by McMahan's moral philosophy on the one hand and positivist legal theory on the other. McMahan's position can be characterized by the argument that we should

> distinguish sharply and explicitly between the morality of war and the law of war. The morality of war is not a product of our devising. It is not manipulable; it is what it is. . . . the laws of war are conventions we design for the purposes of limiting and repairing the breakdown of morality that has led to war, and of mitigating the savagery of war, seeking to bring about outcomes that are more rather than less morally desirable. (McMahan 2008, 35)

McMahan does not believe that the morality of war will conflict with the laws of war very often, but there are crucial areas where it will. Despite the twin recognition that there are prudential reasons to have a legal framework that does not live up to rules of deep morality and that there are moral reasons for having laws that do not live up to the first-order demands of morality, he still argues that "when the morality of war requires what the law forbids . . . one must do what morality requires" (McMahan 2008, 38–39). McMahan believes he shares an account of the basic premises of the morality of war with Walzer and is highly critical of the conclusions Walzer reaches from this premise. This premise is the claim that just war principles are "grounded in claims about the possession and forfeiture of individual moral rights" (McMahan 2008, 21). However, a plausible account of Walzer's moral theory has to take into account his view of how politics and society—the collectivization of moral experience—informs his account of international ethics and just war theory. Walzer's ambition was always "to take a stand among the universalists," but crucially it was to be "a non-standard variety, which encompasses, perhaps even helps to explain the appeal of moral particularism" (Walzer 1990, 509). Walzer's "thin" and "reiteratively learned" universalism is itself an explanation about how a world characterized by moral pluralism develops shared moral and legal rules—most importantly, humanitarian rules proscribing "acts that shock the conscience of mankind" (Walzer 2000, 21). Recognizing the rights of men and women in Walzer's work comes hand in hand with respect for pluralism, otherness, and, most importantly, the recognition that universal rights are thin. They are urgent and in need of protection, but they are not the basis of a fully elaborated morality. That we save for our "thick" social and political projects. Walzer's ideal global constitution and his just war theory is premised on a set of principles that (when given "covering law" formulation) sounds like individual rights but is in fact grounded in the idea of a legalized international society with principles of equality, collective self-determination, and international (legal) human rights as an outer limit to toleration (Walzer 1994, 2004). Walzer's conception of universal rights is legalized in ways that go far beyond conservative conventionalism, but it shares little common ground with analytic cosmopolitanism or revisionism. Crucially, this means that Walzer does believe that killing in war is morally different from killing generally—something that McMahan describes as "implausible" (McMahan 2009, 36). Yet, as Ian Clarke argues, international society frames discussion of war as a distinctive act to which special moral rules apply and "it is not ours to wish away" (Clarke 2015, 19).

Ultimately, I do not think that McMahan and Walzer (and the legalists) are arguing from the same moral premise, and, crucially, their distinct understanding of the nature and purpose of just war theory means that they are engaged in two very different exercises. As such I put this critique of conventionalism to one side. At the other end of the spectrum we have the positivist legal position. As the position only serves as a marker I will render it briefly. Legal positivism argues that the normativity of the laws of war stems from the consensual (conventional and customary) sources of those norms. Justice is a social artifact born of *lex lata* (the law as it is) and moral reasoning is *ultra vires* (beyond legal authority). The justice of a rule does not depend on whether it meets moral criteria. There is a deeper antipathy to morality here, one linked to the realist tradition in International Relations. To admit moral reasoning into this process is to invite chaos as the law specifies how moral pluralism is mediated institutionally. In between these two poles we have the positions of Buchanan and the legalists. On the face of it they share significant common ground. Buchanan's *Justice, Legitimacy and Self-Determination* and all of his work since that point focuses on the moral foundations of international law. His major innovation is that his account recognizes that international justice must engage in what he terms institutional moral reasoning.

Buchanan offers what he terms a progressive conservatism by which he means

> that the theory should build upon, or at least not squarely contradict, the more morally acceptable principles of the existing international legal system. The most obvious reason for this requirement is that satisfying it will generally contribute to the accessibility of the theory's proposals. . . . But there is another reason: Where possible the theorist should build upon the moral strengths of the existing system, because it would be irresponsible to advocate, unnecessarily, a disregard for whatever progress has already been achieved in the system. (Buchanan 2004, 63)

In later work, particularly in *The Heart of Human Rights*, Buchanan argues forcefully that recognizing the institutional and legal character of norms requires a distinct approach to political theory. In the context of an argument about human rights Buchanan argues against what he calls the "mirroring view" that a necessary element of the justification for a norm is the existence of a corresponding, antecedently existing moral right (Buchanan 2013, 14).

There is something (morally) important about the fact that human rights law or the laws of war are institutionalized legal principles. Buchanan's practice-based approach to questions of international law and justice is intended to offer a critical understanding of our highly legalized and institutionalized moral lives. He does not deny that there are independent moral principles (he refers to his basic moral commitment to a natural duty of justice), but he argues that morality has to accept the constraint of moral accessibility, something that signals that nonideal theory should make a case that the corresponding ideal theory's principles can be satisfied or at least seriously approximated through a process that begins with the institutions and culture we now have and that does not involve unacceptable moral wrongdoing in the process of transition. The new centrality of human rights law to international law in general is what gives him the confidence to argue that just war theory is seriously deficient. His core argument is that the legalists do not realize that some plausible institutional reforms could provide the basis for changes to the laws of war that a proper appreciation of the moral implications of human rights demands (Buchanan 2004, 62).

The legalists or conventionalists make up the final point on this scale. In common with Buchanan they are interested in the institutionalized nature of international justice. Walzer contrasts "thin" international morality based on politically (reiteratively) constructed universal principles with "thick" morality at the domestic level (see also Sutch 2009). In a more recent contribution to this tradition Stephen Ratner (echoing Walzer) refers to the "thin justice of international law" (Ratner 2015). Thin, in both of these cases, is not meant to infer lack of substance in international justice but to indicate the distinctly international law based nature of the ideas of justice in play. Despite this I do not think that contemporary just war theorists can be accused either of ignoring moral principles and adopting a conventional approach to the law of war or of institutional blindness. On the contrary some of the best interventions here offer a critical view of the law and show why the ethics of war should always be treated as institutional moral principles. McMahan and Buchanan are correct to note that the institutional context has a significant effect on just war norms, but they do not go far enough. War is an institution embedded in the normative framework of international society. The moral vocabulary shared by agents is shaped by and, in turn, shapes the international legal order. The contemporary just war theorists they criticize are not traditional just war theorists at all. Each of them (and here I use Walzer, Ratner, Christian Reus-Smit, and Andrew Hurrell as examples) has a highly contemporary argument about why ethics must be so closely related to politics and law, and none of them can be rightly accused of blundering blindly into the belief that law and morality simply coincide. Crucially, the legalists give reasons for being more cautious about the transformational impact of international human rights law, and this often makes their prescriptions more conservative than those Buchanan advances. In this case the argument is that there are institutional moral reasons for not developing just war rules that might satisfy Buchanan's human rights requirements. The issue hinges on who gives the best account of what justice requires. I argue that the close association between the normative claims of morality and law is both philosophically and ethically credible and essential to the understanding of the sociolegal normativity of just war claims, and that this falls short of Buchanan's ideal. The critical effect of the increasing normative importance of humanitarian and human rights principles does challenge the existing rules of war in important ways, but that effect is limited by the broader constitutional structure of the global legal order

War, Law, and Ethics

The apparent renaissance of just war theory after 9/11 should be viewed as part of the legalization of world politics rather than a resurgent triumph of ethical reasoning over a conservative legal framework riven with instrumental compromise. Here the critical edge of just war theory appears to be sheathed in its own successful conventionalization—its practices harnessed to the law. However, there are both practical (instrumental and pragmatic) reasons for embracing the institutionalization of just war theory, and, I argue, we gain access to a rich account of the normativity of just war theory (and other forms of practical reasoning in contemporary global affairs) as we do so. Stephen Ratner outlined this space in a recent article in *International Theory*:

> These two disciplines are uniquely competent to advance our understanding of global justice and the means to achieve it. Ethics, in the form of political and moral philosophy, poses the most fundamental questions about responsibilities at the global level and aims to produce a tightly reasoned set of frameworks and principles regarding the possibility of a just world order. International law, with its focus on legal norms and institutional arrangements devised by global actors, provides a path, as well as illuminates the obstacles to implementing theories of the right or of the good; indeed it turns out to have its own internal morality as well. Yet despite the complementarity of these two projects, neither is drawing what it should from the insights of the other. The result is ethical scholarship that often avoids, or even misinterprets, the law; and law that marginalizes ethics even as it recognizes in some sense the importance of justice. (Ratner 2013, 2)

The internal morality of law (something Reus-Smit refers to as the institutional autonomy of law) is an essential part of the equation. The internal morality of the laws of war is rooted in a conception of legitimate norm generation, which is largely based on state consent but has recently developed (or is presently developing) principles that transcend state consent. It is here that we find thin, solidarist accounts of justice that underpin just war doctrine and stronger cosmopolitan claims that urge significant reform. Human rights principles are a key part of this development, but any account of how human rights should alter our moral assessment of the laws of war has to keep the internal morality of this institutional framework in mind.

The legalization of international politics (as the term of art has it) has sets of characteristics that mark out the institutional development of international society as a distinct form of political society—one in which the normative is configured in a unique fashion. The plurality of legal regimes, their fragmented and nonhierarchical relationship (in theory at least) with each other and with international actors, the absence of anything like an orthodox legislature are vitally important aspects of this character. Yet we also see the development of constitution-like characteristics at the international level and the development of substantive normative hierarchies (where certain core norms—including some aspects of human rights and humanitarian law—develop peremptory status or become the basis for rules that are binding irrespective of state consent or the basis for universal jurisdiction). This politically and legally contested space is a rich context for thinking about global justice in a way that connects directly with global actors. In this broad context I am particularly interested in the ways that IPT uses human rights claims as a proxy for (or sometimes simply as a starting point for) moral argument. In much IPT and most explicitly in liberal-cosmopolitan theory the pattern of argument often takes the following form—human rights are universal and this implies greater obligations to eradicate poverty/to defend by force the rights of foreigners/to fight for democratization/to fight preventative wars against international terror. Indeed, there appears to be little that human rights cannot demand of international actors. But human rights come with legal baggage—they are a product of a specific time and place—and that means that while the appeal of human rights based reasoning is clear (in that the theorist can draw on existing universal principles that are the enshrined in treaty and custom and that underpin serious moral debate about the reform of existing institutions and norms), there is also a cost. The further away from the law a theory gets the more difficult it is to claim that its account of human rights is normative. Here I use the word *normative* in a way that reflects the legal sense of the term. For lawyers the normative is binding; it imposes obligations on actors because the rule it evokes is law. In what follows I argue that an appropriate conception of the normative is broader than this, but that it is not possible to sever the link between the normative understood in the legal sense and the normative understood in the moral sense if we wish to harness the power of the legalized world order to the work of IPT.

Cosmopolitan Institutionalism and the Reform of the Laws of War

It is this form of legalized political theory that characterizes the cosmopolitan-institutionalism of Buchanan and Keohane and the conventionalism of Walzer. The basic premise of Buchanan's political theory leads him to argue that "institutionalised moral reasoning" is essential to the justification of ideas of justice and legitimacy. This means both that the theorist ought to begin with the institutions and culture we now have but that in order to retain a critical edge,

> philosophical reasoning is needed to determine what sort of epistemic virtues institutions must have if they are to help determine the content of norms. To put the same point differently, traditional philosophical theorizing about human rights needs to be augmented by social moral epistemology, understood as the systematic comparative evaluation of alternative social institutions and practises as to their effectiveness and efficiency in forming beliefs that are critical for moral judgment and justification. (Buchanan 2004, 5–6)

Buchanan maintains that the just war/legalist position is "methodologically flawed" because it is not empirical enough and that the costs of continued adherence to the just war norm are therefore "intolerably high" (Buchanan 2004, 264). Its conservatism stems from its failure to treat the normative force of human rights with the institutional creativity it demands. The core issue at stake concerns whether the growing normative force of human rights (clearly evidenced by the development of the international legal order since 1945) should license reforms to the laws of war to enable the pursuit of goals including humanitarian protection or the defeat of international terrorism. Buchanan and Keohane claim that conventionalism is uncritically conservative in its respect for existing legal provisions and that situating the laws of war in the context of the international legal order as a whole and having the imagination to see plausible institutional reforms could push us to recognize the legitimacy of legal reform.

The two key elements that unpack Buchanan's argument about institutional moral reasoning are the centrality of human rights to the idea of justice and the idea of institutional reform. The first claim is that there has been a "transition from an international legal system whose constitutive legitimizing aim was peace among states (and before that the regulation of war among states) to one that takes the protection of human rights as one of its central goals" (Buchanan 2010, 72). A conservative approach to this development rests its case on what he terms "the parochialism objection"—the idea that human rights, as expressed in international institutions and norms, are limited sets of values (limited to the capitalist-liberal values of the powerful, for example) or arbitrarily ranked so as to privilege the interpretation of the powerful (Buchanan 2010, 73). However, it is possible to mitigate this risk by instituting institutional reforms that would enable us to work confidently toward the goals indicated by a commitment to human rights.

Buchanan and Keohane have developed arguments to this effect covering both *ad bellum* and *in bello* issues. In "The Preventative Use of Force: A Cosmopolitan Institutional Proposal" they advocate reform of the UNSC to remove the "legitimacy veto" from the P5 (the five permanent members of the Security Council) and to establish *ex-ante* and *ex-post* accountability mechanisms that would provide institutional assurances against the misuse of a much-needed doctrine of preventative self-defense and humanitarian intervention (Buchanan and Keohane 2004). In "Towards a Drone Accountability Regime" they acknowledge the risks posed by the availability of drone technology but propose an international regime that holds states accountable to interstate and transnational institutions as well as domestic institutions that would mitigate the risks of using the technology while enabling states to avail themselves of the attractions of using drones (Buchanan and Keohane 2015, 18). While the specifics of the regimes are worthy of serious attention in both cases Buchanan and Keohane think that the prospect of establishing these accountability regimes at an international level is limited. In the case of UNSC reform, charter provision requiring the acquiescence of the P5 to charter reform rules out the prospect (Buchanan and Keohane 2004, 17). In the case of drone accountability getting major powers to consent to a formal binding regime—and elsewhere Buchanan has called the need for consent as amounting to a legitimacy veto (Buchanan 2010, 111)—rules out the prospect (Buchanan and Keohane 2015, 25). In both cases the imperatives for developing the new institutional frameworks lead the authors to argue for a compromise position that is nonuniversal. In place of a reformed UNSC Buchanan and Keohane argue for a coalition of reasonably democratic states that would implement the accountability mechanisms in order to authorize the legitimate use of force and sanction the illegitimate use of force (Buchanan and Keohane 2004, 20). In the case of the drone accountability regime they propose an "informal interstate arrangement" between "'first-movers'—in this case states that are already using lethal drone technology" (Buchanan and Keohane 2015, 17, 25).

In both cases the refusal of conventionalist just war theory to move beyond a restrictionist position (however fruitlessly in the face of great power behavior) is not the product of a denial of the legitimacy of the idea of preventative self-defense, or of humanitarianism, or of the acceptability of drone technology (legitimately deployed). Rather, the reluctance stems from the risks associated with institutions too weak to ensure conformity with relatively uncontroversial moral and legal norms (Buchanan and Keohane 2015, 23; Buchanan and Keohane 2004, 250–79). However, I think that this claim moves Buchanan and Keohane beyond the scope of their foundational claim of being engaged with the norms and institutions that we currently have. Not only are the arguments they rely on controversial in their interpretation of the normative reach of those norms (especially of human rights norms) but the proposed establishment of nonuniversal institutions that would permit violations of restrictionist norms is highly controversial—reintroducing a distinction between civilized (or legitimate) states and noncivilized (or rogue) states that undermine the contemporary international legal order.

Legalization and IPT

The essence of a political theory of international law is the claim that starting from within the discourses of international society gains the theorist of justice access to what Andrew Hurrell terms "a stable and shared framework for moral, legal and political debate" as well as "a stable institutional framework for the idea of a global moral community" (Hurrell 2007, 303). Accessing and relying on this shared framework "by defining a problem or issue as legal reduces opportunity costs by invoking standardized, socially sanctioned solutions" (Reus-Smit 2005, 38). The key point is that if normative claims are going to rely, in part, on socially sanctioned solutions then significant work must go into critically exploring the socially sanctioned nature of key claims. Crucially,

> Legal claims are legitimate and persuasive only if they are rooted in reasoned argument that creates analogies with past practise, demonstrates congruence with the overall systemic logic of existing law and attends to the contemporary social aspirations of the larger moral fabric of society. (Finnemore and Toope 2006, 195)

This is not to adopt an unduly conservative perspective that is immune to moral revision. Reus-Smit argues that paying attention to the social norms that underpin the constitutional structure of international society underwrites a renewed focus on normative and ethical issues that enables us to move toward a more emancipatory or critical constructivism (Price and Reus-Smit 1998, 259–94). The argument is really about what reasons count as morally or normatively weighty and why and how claims about the moral desirability of reforms to the laws of war interact with claims about a legitimate law-creating process, human rights, and the politics of the use of force. The legalization of debates about the use of force, a process that led to the conventionalization of just war theory, has left an indelible mark on normative debate, but this process creates a new platform for placing normative debate at the heart of the law and politics of warfare. As just war theory became legalized it became conventionalized, losing its moral distance and critical bite. As it reemerged in the post–Cold War era, just war theory (especially in the debates within international institutions rather than in the academy) is both tied to the value structures of the international legal order and provides crucial space for moral debate. Some forms (in my view the most useful forms) of just war theory are now central to debates concerning the development/reform of positive law *precisely because* they have been legalized. But this comes with constraints on how we invoke key norms—including (perhaps especially) international legal human rights norms.

Human Rights and War

My main criticism of the cosmopolitan-institutionalist approach to the subject focuses on the ways that human rights norms are invoked to justify assertive policies. Consider the following claim from Buchanan:

> At the dawn of the modern human-rights era, the role of human rights in the international legal order was rather minimal. . . . The situation is different today. . . . There is growing acceptance of the idea that conformity to human rights norms is a necessary condition of the legitimacy of governments and even of states. . . . these developments signal the transition from an international legal system whose constitutive, legitimizing aim was peace among states (and before that merely the regulation of war among states) to one that takes the protection of human rights as one of its central goals. (Buchanan 2004, 71)

This statement underwrites his commitment to humanitarian intervention, preventative self-defense, remedial rights to secession, the drone accountability regime, and even wars fought to democratize authoritarian states. There is undoubtedly some merit to the claims, but when we explore the detail I think we find that it strains the normative beyond credibility.

In order to explore this we can gain real insights from Theodore Meron's exploration of the "Humanization of International Law." Central to this idea is the claim that the progressive development of humanitarian law is now guided by human rights standards (Meron 2000). This has been a vital element of the expansion of robust legal and institutional responses to gross violations of human rights, which are often characterized as crimes against humanity and war crimes. In particular, it has served as the basis for the justification for humanitarian military action, the erosion of immunity for criminal culpability for violations of the laws of war, and the establishment of ad hoc and permanent international criminal tribunals.

A second vital claim is that the gradual legal development of "principles of humanity and of the public conscience" that is found in all human rights conventions, from the United Nations Declaration of Human Rights to the Rome Statute, is creating new normative hierarchies in public international law (Meron 1986; Thirlaway 2003, 142; Pustogarov 1999, 134; Chetail 2003, 338). The Martens Clause (named after the diplomat who inserted the "acts that shock" clause into the preamble of the Hague Convention) is taken to be part of customary international law and is therefore of universal scope. It serves to remind all actors that there are some humanitarian standards that are binding regardless of whether a state has become party to a particular treaty or has reservations against some elements of treaty law (Cassese 2000, 192). The influence of human rights has been central to this process.

As early as 1950 human rights principles were being used to flesh out the meaning of the Martens Clause (Cassese 2000, 207). Now, however, human rights principles are being used to assert the peremptory nature of norms closely associated with "acts that shock" in a way that has started "a limited transition from bilateral legal relations to a system based on community interests and objective normative relationships" (Meron 1986, 256). It is the development of this hierarchy of norms that suggests to some the subversion of the state-consent model of international law and its replacement with a hierarchy of universally binding human rights–based norms.

It is worth remembering that Walzer's just war theory, from *Just and Unjust Wars* to *Arguing about War*, is justified in reference to "acts that shock." Walzer describes his response to such acts—collectively recognized as outrages by the reiterated, universal, but thin moral sense that underpins humanitarianism—as exceptions to the general rule of nonintervention. It is also the case that Walzer explores the idea of institutional reform in the light of this shared morality in the final essay of *Arguing about War* entitled "Governing the Globe." The simple fact that Walzer proposes exceptions and considers institutional reforms defeats the basic claim that he simply subordinates morality to the law. However, the real issue between conventionalist and cosmopolitan just war theory concerns the extent to which this hierarchy of norms, itself driven by the moral value of human rights, grants license to alter the norms and institutions that govern conflict.

In law the issue is not clearly resolved. Norms associated with war are often associated with hierarchical provisions. For example, the prohibition of genocide, slavery, torture, and aggressive war are, because of their normative urgency, protected by laws of a special character (they are binding irrespective of consent and nonderogable) and give rise to community obligations (even without a direct link to—or nexus with—the specific violation). However, while the categories of norms *erga omnes* (universal obligations) and *jus cogens* (peremptory or compelling law) have been recognized in courts, in treaties, in the declaratory statements of key actors, and in much scholarly work, the idea that we have an uncontroversial hierarchy of norms is still hotly disputed. The way these categories respond to "community interests" and issues of global public policy based on "basic human rights" is fascinating. However, while it is clear that *erga omnes* obligations are associated with "the basic rights of the human person" (*Barcelona Traction Case*), states have shown a real reluctance to test the extent of this concept. While *jus cogens* have clear peremptory status the question of what follows from this in law does not really suggest that these categories fundamentally challenge the norms protecting sovereignty.

There is clear evidence to suggest that human rights norms have become increasingly important to the development of the international legal order. But the bolder claim, that human rights have come to challenge sovereign norms directly, does not have much credibility in legal or social terms. It is certainly not the case that the progressive development of international law leads inexorably to cosmopolitan conclusions. A clear example can be found in the 2006 judgment of the International Court of Justice in *Armed Activities on the Territory of the Congo (Democratic Republic of Congo v. Rwanda)* (Armed Activities Case). In this judgment the court was considering whether it had the jurisdiction to hear a contentious case between the two parties. The case centered on "massive, serious and flagrant violations of human rights and international humanitarian law . . . perpetrated by Rwanda on the territory of the DRC." In its judgment the Court began by

> reaffirming that "the principles underlying the (Genocide) Convention are principles which are recognized by civilized nations as binding on States, even without conventional obligation" and that a consequence of that conception is "the universal character both of the condemnation of genocide and of the co-operation required 'in order to liberate mankind from such an odious scourge' (Preamble to the Convention)." (at paragraph 64)

The *jus cogens* character of the prohibition on genocide gives voice to the centrality of humanitarian principles at the heart of contemporary international law. Nevertheless, the court ultimately found that it had no jurisdiction to hear the case—the *jus cogens* character of a norm and the rule of consent to jurisdiction are two different things (paragraph 64), and no norm of general international law requiring a state to consent to the jurisdiction of the court when dealing with violations of peremptory norms or community obligations exists (see also Shelton 2006, 306–7).

The increasing use of hierarchical normative language, which says something important about the character of humanitarian law, poses hard moral and legal questions. Some international judges are prepared to push the boundaries of the meaning of "acts that shock." Here the views of Judges Christopher Weeramantry and Mohammed Shahabuddeen (in the *Nuclear Weapons Advisory Opinion*) that the threat or use of nuclear weapons could not satisfy the requirements of the Martens Clause are especially important. However, the law, for now at least, is still very respectful of sovereignty and state consent. The opinion of the court in *Nuclear Weapons* was that there existed no customary or conventional law prohibiting the threat or use of nuclear weapons as such, and that, while the threat or use of nuclear weapons would generally be contrary to the rules of law applicable in armed conflict, the court could not conclude definitively whether that would be the case in all circumstances (*Nuclear Weapons*, 265). The issue is not really about the status of human rights, it is about the appropriate way to deal with violations. There is no evidence that cosmopolitan remedial structures (such as the compulsory jurisdiction of courts, unilateral rights of military intervention or secession, declaring nuclear weapons to be illegal without a multilateral convention, and so forth) have any significant support in practice. These difficult legal questions fuel the contemporary just war debate.

Both the conventionalists and the cosmopolitans are asking what we *ought* to do in in response to what we agree are violations of the most morally urgent norms. Both Walzer and Buchanan seek exception to restrictionist rules to deal with "acts that shock." As Buchanan argues forcefully, the relative strength of institutions to deal fairly with hegemonic influence is crucial. Conventionalists adopt pluralist approaches to institutional reform to create political defenses against the imposition of parochial interpretations of communal norms. But state actors adopt pluralist approaches to defend parochial impositions on their own pluralist political communities and to resist intervention. Some actors pursue nonhumanitarian goals under the guise of legitimate interventionism. How are we to decide between the conventionalist and cosmopolitan positions?

The position adopted by Buchanan and Keohane is that the answer is institutional reforms establishing *ex-ante* and *ex-post* oversight over the counter-restrictionist application of *ad bellum* and *in bello* rules required by our normative commitment to protecting human rights. We are morally required to reform the Security Council to enable a dependable and nonparochial response to "acts that shock." The normative force of the need to protect human rights is such that where multilateral institutions are unresponsive to this demand they should be bypassed and supplanted by a league of rights-respecting states or a community of drone-using states. Walzer is also critical of the limitations of the international system (Walzer 2004, 179–82) and believes that the moral force of international human rights requires institutional reform. However, in what he calls his globalist, postmodern project (Walzer 2004, 186), he advocates a dedicated UN military force, a world court, and a new layer of governmental organization but one that sacrifices some opportunities for action on behalf of human rights to minimize the risk of global tyranny (Walzer 2004, 187–88). In other words, Walzer agrees with the moral logic of the cosmopolitan position but balks at the idea that establishing hierarchies in international affairs—goodies and baddies, civilized and uncivilized—is a morally appropriate solution. Given that the moral project is linked to the internal morality of international law (rather than an a priori commitment to cosmopolitanism or communitarianism) the conventionalist position seems to be the stronger because there is no evidence in the human rights or humanitarian legal order that there is a moral basis for the creation of legal hierarchies between states.

The argument that we need a weakened prohibition on the use of force and more license to use force than that granted in international law because we have morally urgent reasons to prevent humanitarian atrocity or defeat terrorism is entirely defensible. It is equally clear that in going down this road we risk permitting the use of force in ways that cause the deaths and suffering of innocent people. The idea that we might establish institutions to mitigate that risk is eminently sensible, particularly as the corresponding argument is that, in the absence of these safety nets, counterrestrictionist policies are not warranted. However, the argument of Buchanan and Keohane, that if we cannot achieve these institutional safeguards multilaterally we ought to set up less-than-universal institutions, cuts to the heart of international legal order—the principle of sovereign equality.

One of the principal advances in contemporary international law is the eradication of the distinction between civilized and noncivilized peoples and the establishment of what Reus-Smit calls an equalitarian regime (Reus-Smit 2005). This equality, protected by the rules of noninterference, self-determination, and state consent, is to be favorably contrasted with the indignity of the imperial and colonial period where uncivilized peoples fell outside the ambit of the law and could become subject to "civilized" peoples. The equalitarian regime is essential to the recognition of the inherent dignity of all peoples, and the thought that the right to decide upon the right to use force might settle on a self-identifying group of rights-respecting, democratic states—still less a group of states whose sole criteria for membership is that they have lethal drone technology—seems to place far too much weight on the extent to which the international community prioritizes human rights norms over sovereign equality.

The universal recognition that there are acts that shock the conscience of mankind that are prohibited and that the international community has obligations to do *something* about it does not settle the argument about what to do about it. It certainly does not in itself warrant the conclusion that respect for human rights requires the compulsory jurisdiction of courts, the right to kill enemies in the sovereign territory of nonconsenting states, or the right to treat some enemy combatants as not having Geneva rights. It does not, in my view, warrant the weakening of the general prohibition on the use of force in ways that make fighting terror easier but make considerations of last resort or the discrimination between combatant and civilian less easy.

Conclusion

The reason international law prohibits the unilateral use of force or torture, or restricts the right to use drones in the territory of a nonconsenting state, is because that is what states have agreed to as the best way to protect the spaces in which people give concrete meaning to human rights. We are very aware that affording peoples this space to be self-determining makes it difficult to always defeat oppression and terror. But loosening those restrictions risks creating precisely the hierarchy that contemporary international law seeks to avoid. It risks, in Walzer's words, tyranny. The normative force of human rights gives us a place to stand to criticize the international order and seek reform but not to use force to pursue human rights reforms. It does give us license to engage with "acts that shock," but successive attempts to negotiate a complete solution (such as the attempt to develop the Responsibility to Protect) have not garnered support for really changing the law. In part at least this is because international humanitarian law is not intended to deal with what Walzer has called "ordinary oppression." Rather, it is intended to deal with "extreme oppression." Terry Nardin makes Walzer's position clear:

> To intervene in situations in which the abuses do not "shock the conscience of mankind" would be to improperly infringe the independence to which states are morally entitled, within wide limits. It may be hard to draw a line between great and little crimes, but not that hard: we are *not* looking at a continuum here, Walzer insists, but at "a chasm, with nastiness on one side and genocide on the other." (Nardin 2013, 73)

Without consensus on what "shocks the conscience of mankind" *and* on what to do about it, the restrictions remain in place. Action taken in violation of those restrictions may be "illegal yet legitimate" and the world brought to see the power of the mitigating argument. But the world does not see a clear human rights basis to delegate the authority to weaken existing *ad bellum* and *in bello* rules to a small group of states (no matter how civilized).

Cases

Legality of the Threat or Use of Nuclear Weapons, Advisory Opinion, ICJ Reports 1996, p. 226.

Armed Activities on the Territory of Congo (New Application: 2002) (Democratic Republic of the Congo v, Rwanda), Jurisdiction and Admissibility, Judgment, ICJ Reports 2006, p. 6.

Barcelona Traction, Light and Power Company, Limited, Judgment, ICJ Reports 1970, p. 3.

References

Beitz, C. 2009. *The Idea of Human Rights*. Oxford: Oxford University Press.

Brunée, J., and S. Toope. 2004. "Slouching towards New 'Just Wars': The Hegemon after September 11th." *International Relations* 18(4): 405–23.

Buchanan, A. 2004. *Justice, Legitimacy and Self-Determination: Moral Foundations for International Law*. Oxford: Oxford University Press.

Buchanan, A. 2010. *Human Rights, Legitimacy and the Use of Force*. Oxford: Oxford University Press.

Buchanan, A. 2013. *The Heart of Human Rights*. Oxford: Oxford University Press.

Buchanan, A., and R. Keohane. 2004. "The Preventative Use of Force: A Cosmopolitan Institutional Proposal." *Ethics and International Affairs* 18(1): 1–22.

Buchanan, A., and R. Keohane. 2015. "Toward a Drone Accountability Regime." *Ethics and International Affairs* 29(2): 15–37.

Cassese, A. 2000. "The Martens Clause in International Law: Half a Loaf or Simply Pie in the Sky?" *European Journal of International Law* 11(1): 187–216.

Chetail, V. 2003. "The Contribution of the International Court of Justice to International Humanitarian Law." *International Review of the Red Cross* 85: 235–69.

Clarke, I. 2015. *Waging War: A New Philosophical Introduction*. 2nd ed. Oxford: Oxford University Press.

Doyle, M. 2008. *Striking First: Prevention and Preemption in International Conflict*. Edited by S. Macedo. Princeton, NJ: Princeton University Press.

Egede, E., and P. Sutch. 2013. *The Politics of International Law and International Justice*. Edinburgh: Edinburgh University Press.

Finnemore, M., and S. Toope. 2006. "Alternatives to 'Legalization': A Richer View of Law and Politics." In *International Law and International Relations*, edited by B. Simmons and R. Steinberg, 188–204. Cambridge: Cambridge University Press.

Frost, M. 1996. *Ethics and International Relations: A Constitutive Theory*. Cambridge: Cambridge University Press.

Ginbar, Y. 2008. *Why Not Torture Terrorists? Moral, Practical and Legal Aspects of the 'Ticking Bomb' Justification for Torture*. Oxford: Oxford University Press.

Hurrell, A. 2007. *On Global Order: The Constitution of International Society*. Oxford: Oxford University Press.

International Commission of Jurists. 2009. *Assessing Damage, Urging Action: Report of the Eminent Jurists Panel on Terrorism, Counter-Terrorism and Human Rights*. https://www.icj.org/wp-content/uploads/2012/04/Report-on-Terrorism-Counter-terrorism-and-Human-Rights-Eminent-Jurists-Panel-on-Terrorism-series-2009.pdf.

May, L. 2008. *Aggression and Crimes against Peace*. New York: Cambridge University Press.

McMahan, J. 2008. "The Morality of War and the Law of War." In *Just and Unjust Warriors: The Moral and Legal Status of Soldiers*, edited by D. Rodin and H. Shue, 19–43. Oxford: Oxford University Press.

McMahan, J. 2009. *Killing in War*. Oxford: Oxford University Press.

McMahan, J. 2010. "Pacifism and Moral Theory." *Diametros* 23: 44–68.

Meron, T. 1986. "On a Hierarchy of Human Rights." *American Journal of International Law* 80(1): 1–23.

Meron, T. 2000. "The Humanization of Humanitarian Law." *American Journal of International Law* 94(2): 239–78.

Nardin, T. 2013. "From Right to Intervene to Duty to Protect: Michael Walzer on Humanitarian Intervention." *European Journal of International Law* 24(1): 67–82.

Price, R., and C. Reus-Smit. 1998. "Dangerous Liaisons? Critical International Theory and Constructivism." *European Journal of International Relations* 4(3): 259–94.

Pustogarov, V. V. 1999. "The Martens Clause in International Law." *Journal of the History of International Law* 1: 125–35.

Ratner, S. 2013. "Ethics and International Law: Integrating the Global Justice Project(s)." *International Theory* 5(1): 1–34.

Ratner, S. 2015. *The Thin Justice of International Law: A Moral Reckoning of the Law of Nations*. Oxford: Oxford University Press.

Reus-Smit, C. 2005. "Liberal Hierarchy and the Licence to Use Force." *Review of International Studies* 31: 71–92.

Rodin, D. 2002. *War and Self-Defense*. Oxford: Oxford University Press.

Rodin, D., and H. Shue. 2008. *Just and Unjust Warriors: The Moral and Legal Status of Soldiers*. Oxford: Oxford University Press.

Sands, P. 2008. *Torture Team: Deception, Cruelty and the Compromise of Law*. London: Penguin.

Shelton, D. 2006. "Normative Hierarchy in International Law." *American Journal of International Law* 100(2): 291–323.

Sutch, P. 2009. "International Justice and the Reform of Global Governance: A Reconsideration of Michael Walzer's International Political Theory." *Review of International Studies* 35(3): 513–30.

Thirlaway, S. 2003. "The Sources of International Law." In *International Law*, edited by M. Evans. Oxford: Oxford University Press.

Waldron, J. 2018. "Deep Morality and the Laws of War." In *The Oxford Handbook of Ethics of War*, edited by S. Lazar and H. Frowe. Oxford: Oxford University Press.

Walzer, M. (1977) 2000. *Just and Unjust Wars: A Moral Argument with Historical Illustrations*. 3rd ed. New York: Basic Books.

Walzer, M. 1990. "Nation and Universe." In *Tanner Lectures on Human Values XI*, edited by G. Peterson. Salt Lake City: University of Utah Press.

Walzer, M. 1994. *Thick and Thin: Moral Argument at Home and Abroad*. Notre Dame, IN: University of Notre Dame Press.

Walzer, M. 2004. *Arguing about War*. New Haven, CT: Yale University Press.

Chapter 2

The Fantasy of Nonviolence and the End (?) of Just War

LAURA SJOBERG

We usually talk about war as if there is peace. We usually talk about conflict as if there is a such thing as an absence of conflict. We usually talk about violence as if there is such a thing as nonviolence. These dichotomous ways of thinking have in common that they define, understand, process, and evaluate the perceived evils of war, conflict, and violence by using an idealized understanding that there is a possible world without them. We often process the evils of war and violence with the hope and ideal that there is a pure alternative to strive toward.

This dichotomous thinking is complicated by, but evident in, just war theorizing. In just war theorizing, the dichotomy becomes a trichotomy that still relies on dichotomous thinking: there is unjust war (the ultimate evil); just war (necessary evil and morally permissible); and just peace (that toward which just war strives). There are several basic assumptions in this trichotomy that not only just war theorists but also many scholars, politicians, and even everyday people take for granted. The first assumption is that there are distinguishable moments that can tell onlookers when war starts, what war is, and when war ends—war and violence are things that are time delineated; they start and end, approximately if not precisely. The second assumption, which is required for the first, is that the war/peace and violence/nonviolence dichotomies make sense. The unspoken idea behind this is that war can be transformed not only into not-war but into peace. The third assumption is that there is an evil—war—that can be not only morally acceptable but the right thing to do. There is bad violence, acceptable/good violence, and nonviolence (the cause that acceptable/good violence serves).

These are not assumptions that are subject to a significant amount of critical examination either in just war discourses or in everyday practices of politics or ethics, even if some scholars and some scholarship question some elements of them. They are, instead, often taken largely for granted. Most histories of wars, most news reporting about wars, most ethical analyses of wars—nearly all those discourses share these assumptions, either implicitly or explicitly.

This chapter makes the argument that the assumption that war and peace exist conceptually or empirically, or both, is deeply problematic. Building on feminist theorizing about the links between sexism, patriarchy, and violence and the continua of violence, this chapter argues that there is no nonviolent alternative to violence. After laying out its theoretical approach to violence, this chapter turns to exploring that interpretation's implications for just war theorizing. It contends that there is no additive or multiplicative approach for *jus ad bellum*, *jus in bello*, and *jus post bellum* that can account for thinking about violence as a continuum. It argues that a continuum approach to violence has a number of important implications for many of just war theorizing's concepts, as well as for the overall utility of just war thinking. The chapter concludes by exploring the implications of this approach for everyday thinking about the practices of violence.

A Continuum Approach to Violence

In 1985, Betty Reardon's *Sexism and the War System* made the argument that sexism and militarism are not separate or separable phenomena but intrinsically linked. Reardon (1985, 5) argues that "sexism and the war system are two interdependent manifestations of the same problem: social violence." She explains that the term *war system* is not meant to be a synonym of or a stand-in for the idea of war, but instead a transgressive reinterpretation: "My use of the term *war system* refers to our competitive social order, which . . . assumes unequal value among and between human beings, and is held in place by coercive force" (Reardon 1985, 10, emphasis in the original). Reardon saw the war system as manifest everywhere, not only in places where active state-to-state military conflict was taking place. In Reardon's (1985, 11) view, "the war system pervades our lives and affects every aspect of society from the structural to the interpersonal."

Reardon (1985, 13) sees war, warfare, militarism, and militarization as subconcepts within the war system. She contends that patriarchy is a key principle structuring the war system and its elements, where "the militarization of society is the unchecked manifestation of patriarchy as the *overt* and *explicit* mode of governance" (Reardon 1985, 16, emphasis in the original). She contends that patriarchy—political and social investment in the dominance of men and masculinities in social, political, and economic structures and spaces—is both constitutive of and constituted by the war system.

Reardon links the interdependent relationships between sexisms and violence to a feminist continuum approach to violence. Others have also described violence as a continuum always and already implicated in sex discrimination (e.g., Cuomo 1996; Cockburn 2010; Moser and Clark 2001; Kelly 1987). In my reading of this work, I see a feminist continuum approach to violence as containing four inherent principles. First, war is not just an event (e.g., Cuomo 1996; Wibben 2010; Pain 2015). Many people assume that war is a historical event that starts on some given day, proceeds through various battles and strategic and tactical encounters, and ends on some given day on which a cease-fire is declared or a treaty is signed. This view is taught in schools and often used in history books. Both the timeline element of this assumption and the concept element are wrong, and contradicted by a nonevent approach to war. The timeline element neglects the fact that, even were one to accept some traditional idea about what war *is*, it cannot be neatly packaged in the ways that politicians and politicized historians often try to sell it.

Commonly used definitions of war range from very narrow to broader interpretations. Very narrow definitions require war to be between recognized and structured nation-states, openly declared by state governments, fought on traditional battlefield in uniforms, with a minimum number of battlefield casualties (Vasquez 2009, 22; Russett 1994). Others recognize that these requirements are both outdated and unrepresentative (Kaldor 1999). Wars (even those between states) are now rarely if ever declared (Fazal 2012). Wars are fought largely away from traditional battlefields, and many fighters neither fight proximate to each other nor in uniforms (e.g., Megret 2011; Pfanner 2004). Calculating battle deaths is tricky and political (e.g., Lacina and Gleditsch

2005). Above and beyond that, many if not most wars are within rather than between recognized nation-states, involving paramilitary organizations, gangs, rebels, and state militaries, sometimes in incomprehensible combinations (e.g., Fearon and Laitin 2003; Lake 2003). Whether war is defined in "old" or "new" ways, war historians and political scientists often define wars' trajectories by beginnings and ends, even engaging in high-stakes debates about when wars begin and end, and how they are to be understood in relation to each other.

One key "debate" is the relationship between the conflicts known as World War I and World War II. Traditional accounts suggest that World War I started with the assassination of Archduke Francis Ferdinand in 1914, and ended either with the ceasefire in 1918 or with the Treaty of Versailles in 1919 (e.g., Strachan 2003). Similar accounts suggest that World War II started in 1939, with the invasion of Poland by Nazi Germany, and ended in May 1945 in Berlin with the Soviet invasion and in September 1945 in Japan with the American atomic bombing and Japanese surrender (e.g., Weinberg 1995). Other accounts call the two "events" the Thirty Years' War (with reference to the earlier thirty years' war in seventeenth-century Germany) (e.g., Shaw and Westwell 2000). The many sides of these debates share an interest in delineating a start point and an end point.

Yet even in this well-known and well-documented set of conflicts the question of whether it is possible to declare a start point and an end point is not easily answered. "Beginnings" have lead-ups, "ends" have follow-ups, and those do not extend to days or weeks but to months, years, and even decades. Remnants of a part of World War II remain in neo-Nazi movements around the world (e.g., Lemon 2018; Knight 2018). Remnants of World War I can be seen in both the fabric of contemporary political organization and the reverberations of the League of Nations mandate system on postcolonial politics (e.g., Fussell 1975; Winter 1998). The timelines where wars begin and end are themselves social constructions, ideal types, and fantasies.

But the problem with this beginning-and-end notion of war is not limited to the impossible ideal of the timeline of war(s) and conflict(s). Seeing war as an event is a fundamental misinterpretation of what happens in war(s), however defined. If war was *an* event—something that starts and stops—then it would have to be the case that *each* war is neither intimately interconnected with other wars nor connected by similar logics of politics and militarism. Even some traditional war theorists have recognized the impossibility of this assumption (e.g., Vasquez 2009). For example, Carl von Clausewitz, a military strategic theorist often cited by realist security theorists and military planners alike, argued that wars never start—they are simply an extension of politics by other means (Clausewitz 1989 [1832]). Clausewitz also argued separately that wars never end—cease-fires are simply time that the side that is losing the war uses to rebuild its strength to redouble its efforts (Clausewitz 1989). Rather than being an event to Clausewitz (1989), war is a natural and inevitable part of the operation of political systems.

Leaving aside for now the questions of naturalness and inevitability, feminist theorists have thought seriously about, and critiqued heavily, the notion that war is an event (e.g., Cuomo 1996). Like Betty Reardon (1985), Chris Cuomo (1996) argued that war is present, if sometimes latent, in many if not all of the ways in which everyday life is experienced. Significant documentation shows war to be a multifaceted and amorphous collection of political, economic, social, and military occurrences, experienced in an extended timeframe (e.g., Ghobarah, Huth, and Russett 2003). For example, Cynthia Enloe (2010) traced the life of a mother of a veteran of the 2003 US invasion of Iraq, showing how the soldier's injuries were lifelong, and both the treatment and the pain reverberated not only through the family but through the community. The mother's life, along with those around her, was forever changed by trauma, by support labor, and by financial hardship. Another example can be found in Zainab Salbi's (2006) description of growing up in Saddam Hussein's Iraq, which details the experience of being an Iraqi abroad during the Gulf War, and how living the conflict affected both her personal and professional life in profound and varied ways, and describes loss, confusion, pain, and also a call to service.

The second inherent principle is that violence is a continuum. I do *not* mean to argue that all violence is equal, whether read morally (how evil is it?), materially (how many people were hurt how badly?), or consequentially (what were the impacts of the violence?). The suggestion that a continuum approach to violence likens genocide and small-scale domestic violence is *true*, but not in a sense of equivalence. Rather than suggesting that one spouse hitting another spouse and mass ethnic cleansing have the same moral content, material costs, or consequences, I argue each has common roots, common logics, and possibly common permissive causes.

Take for example Rachel Pain's (2015) work on intimate warfare, or Caron Gentry's (2015) work on everyday terrorism. Neither suggests that household violence should be addressed, either ethically or practically, in the same way as an invasion of a country or the hijacking of an airplane. What both argue, however, is that the logics of these things share at least as much as the points on which they diverge. Like those acts of violence understood as warfare (in Pain's work) and terrorism (in Gentry's work), the act of violence understood as domestic violence operates on the use of violence to enforce fear with the goal of exerting will or control. While this is an overly general and catch-all understanding of the motives and means of acts of violence understood as war and terrorism, it serves as an example. Each exceptional narrative or story that differs from this overly simple account can be matched with a similar exceptional narrative about acts of violence understood as domestic violence. They are not *the same act*, but they are acts of the same *genus*.

What that genus, violence, has in common is three things. First, and perhaps most obviously, they include the use of force. Force here is and should be broadly defined to include physical violence, psychological violence, and epistemic violence (e.g., Moreton-Robinson 2011), among other potential types. Again, this is not to hold equivalence between either the types of violence or particular violences within each category—it is only to say that all kinds of violences exist along a continuum of violences. Violences' second commonality is that the use of force is unified not by one logic but by an interconnected set of logics, including logics of coercion, control, annihilation, humiliation, and domination. This is not to say that I can tell you (or anyone) *the* logic of or *the* cause of a given occurrence of violence—such a question might not even make sense (e.g., Muro-Ruiz 2002). Instead, it is to suggest that the (multiple and overlapping) causes of and logics of violence are interrelated. Third, the thread tying together the interrelation between violences is gender subordination.

Scholars who have described this mechanism have used a variety of terms. Reardon (1985) called it sexism; Enloe (1983, 1989) called it patriarchy; Tickner (1992, 2001) called it masculinism. I have previously used the term *gender subordination*, describing it in terms of placing a higher value on traits, objects, persons, and other political entities or groups associated with masculinities over those traits, objects, persons, and other political entities or groups associated with femininities (Sjoberg 2013). Significations that associate things with masculinities (called masculinization), then, are significations of value; significations that associate things with femininities (called feminization) are significations of less, or lack, of value. Along the continuum of violence, it is *not* only men who commit acts of violence, and it is *not* only women who are victims of acts of violence (e.g., Moser and Clark 2001; Sjoberg and Gentry 2007). Women commit violence and men can be its victims, both in militaries and outside of them. But violence, from the bedroom to the traditional battlefield, is gendered and sexualized, where committing acts of violence, especially valorized violence like just war, is associated with masculinity and masculinization (Cockburn 2010; Elshtain 1987; Sjoberg 2013). Victimization, on the other hand, is associated with femininity, feminization, and (inextricably relatedly) devalorization (Peterson 2010).

The third inherent principle, and perhaps the most controversial, is that there is no such thing as peace or nonviolence. This constitutive other category that makes possible the dominant discourses about war and violence is itself impossible. Just war discourses treat war/violence as a "last resort," imagining a (perhaps infinite) set of nonviolent alternatives that ought to be "resorted to" before violence can be justified. The often-stated and sometimes implicit but usually present goal of a just war is the idea of just peace (e.g., Chesterman 2001; Williams and Caldwell 2006)—a postwar order in which justice is served in a way that will prevent the renewal of the fighting and post-hoc justify the utilization of violence. Many popular discourses about war specifically or even politics generally name peace as an aspirational goal, and even describe an ideal of positive peace (e.g., Galtung 1969; Sharp 2012; Brock-Utne 1985). Positive peace in the literature is contrasted with negative peace, where negative peace is the absence of violence or war, and positive peace is associated with justice, fairness, and reconciliation (e.g., Galtung 1969; Kacowicz 1997; Diehl 2016). I argue that neither negative peace nor positive peace are fully possible, and that ignoring their impossibility causes both conceptual and practical problems. In making the argument that there is no such thing as nonviolence or peace, I mean no disrespect to a significant body of literature in peace and peace studies—I simply think it talks about an approximation that many of its progenitors see as harmless, and I see problems with the idea of the approximation.

Particularly, I think that the assumption or the articulation, or both, of the possibility of nonviolent alternatives to war/violence obscures the (perhaps lesser but nonetheless present) violence of these preferred supposedly nonviolent alternatives. An easy example is the United States' discourses in the 1990s praising the economic sanctions regime imposed on

Iraq as a "peaceful alternative" to war (see Arya 2008). Several problems with that classification pervaded the sanctions regime. First, attention to the "nonviolent" sanctions regime distracted attention from significant simultaneous bombing campaigns, alternatively described as accompanying or enforcing the supposed always-existing peace (see Gibbons 2016). The bombing campaigns became part of a "peaceful" solution, despite being of a severity matched very few times in history (see Shah 2002). Absent the sanctions regime's discourse of peace, the accompanying bombing may very well have been described as a war (see Arkin 1999). With the sanctions regime, it was characterized as violence in support of co-existing nonviolence. In my view, this is an oxymoron. Second, the sanctions regime itself was full of violence, even if that violence was not hand-to-hand combat on a battlefield. The humanitarian consequences of the embargo on Iraq from 1991 to 2003 are well documented—between half a million and a million Iraqis died of starvation, malnutrition, and denial of medication and vitamins over that period in a country that had previously been middle class (Alnasrawi 2001). Some critics went so far as to call the sanctions regime a genocide (Simons 1999; Gordon 2002). There is substantial disagreement about whose "fault" the sanctions regime was (e.g., Mueller and Mueller 1999)—but the violence inherent therein existed whether the sanctions regime could be fully blamed on Saddam Hussein or fully blamed on the United States, or somewhere in between. Third, the violence of the sanctions regime became intertwined with a wide variety of other, related violences. Putting the starvations and bombings aside, the sanctions regime was amplified by and amplified social unrest, civil strife, and street crime (Gordon 2010). In the sanctions era, the celebrated nonviolent alternative was indeed violent (some say more violent than the characterized "war" that followed), and no actor in the political conflict refrained from the engagement in violence, despite claims to the contrary.

Still, perhaps this example is too easy—because one way to account for it is to suggest that it is evidence of mislabeling of sanctions as nonviolent, rather than evidence of the nonexistence of nonviolent policy options. The hard examples show the same things, though. Argumentation can seem nonviolent, but can include coercion, manipulation, oppression, silencing, and other violences. I argue that what distinguishes the extreme example of sanctions from less extreme examples like argumentation is a difference in degree rather than a difference in type. Violences differ significantly in their types and in their severity. There are physical violences, emotional and psychological violences, discursive violences, epistemic violences, violences of exclusion, violences of poverty, violences of hunger, violences of inclusion—too many types of violence to elaborate here (or perhaps anywhere).[1] I argue that some sort of violence can be found in every claim, every act, and every piece of research, and that it is important to recognize the ever-presence of violence, even when and as violences differ in type or sort or severity.

For example, there is discursive violence in a claim for gender equality that does not recognize the differences among women that constitute in them both potentially different needs and potentially different possibilities to benefit (or not) from policies that claim to be distributing equality (e.g., Butler 1998). The use of the category of "woman" may itself be violent (e.g., Butler 2008). Blindness to class, race, disability, or nationality can be violent (e.g., Alexander 2012). What I left out of and put into the last sentence is itself a violent choice, even when/if I do not fully recognizes its violences. These blindnesses or omissions or statements can be violent not only discursively but also materially, where even claims or politics that are helpful to some (or even most) have violent effects for others. There is violence in the educational and class privilege that produces this chapter (e.g., Clegg 2013), and in the intersection of a wide variety of privileges that produces it in my voice and my voice only (e.g., Frankenburg 1993). That violence may (or may not) be less than the violence that may be inherent in these things never being said, much like the violence in narrow claims to gender equality may or may not be outweighed by the benefit of such a claim for decreasing gender violences. My point (for now) is not to compare the relative social, moral, and political value of particular acts of violence. It is instead to say that there are violences in moves to combat violence, in scholarly discussions of violence, and everywhere.

I would not make this controversial and itself problematic claim if I did not find it important to uncover and stop the obscuring of the violences in actions that are usually characterized as nonviolent. I find it important for a number of reasons. First, I, with many different scholars, am interested in the question of *cui bono* (who benefits) from particular events and behaviors (e.g., Strange 1994; Lundahi 1989; Sjoberg and Gentry 2007). Along these lines, I find it crucial to ask who benefits from any given assumed understanding or framing of a particular concept or political event as nonviolent or peaceful. Relatedly, attention to who is harmed when a concept, understanding of a situation, statement, or policy choice is made is important. Almost every approach to the study of politics suggests that any given policy choice or conceptual framing has winners and losers—making choices implicitly and explicitly is about choosing among potential configurations of winners and losers (e.g., Shafer 1994). Winners experience gains and losers experience violences; sometimes each experiences each if to different degrees.[2] My inquiries here are not about making no losers in any given interaction, or finding a way for the possibility of no violence toward either the winners or the losers. I think that is impossible. Instead, I am interested in naming and recognizing the inevitable violence in any place, situation, choice, or piece of scholarship, and thinking about what responsibility for that means. Third, there is a righteousness to claims of nonviolence or peace, both within just war rhetorics and more generally (e.g., Cahill 2006). While violences are (and should be) morally distinguishable (e.g., McMahan 1993), the righteousness in claims to nonviolence results both in a hubris in the presentation of the claims and in a sense of moral superiority in carrying out the policy actions (see, e.g., Miller 1986).

The fourth inherent principle is that violence along the continuum is distinguishable. The continuum of violence is not a linear progression, where acts of violence can be lined up by severity in a clear way, and severity can be matched one-to-one with moral value (or moral condemnation) of the particular acts of violence. Some comparisons of severity *are* simple (detonating a nuclear weapon in a crowded urban area is infinitely more severe than uttering an insult), and some comparisons of morality *are* simple (killing as an act of genocide is more morally problematic than killing as an act of self-defense).[3] But most comparisons are much more complicated, where they must evaluate differential claims to value in terms of who benefits/who is harmed, what types of violence are used, how severity is understood, and how severity relates to morality. My goal (for now) is not to solve all of these difficult quandaries, but instead simply to suggest that they must be engaged. Often, moral evaluations in war theorizing propose or rely on an ideal or even concrete supposed nonviolence. Recognizing that these "pure" alternatives or goals are straw men complicates moral analyses of violences. Finding no one and no alternative fully nonviolent renders more difficult (and, I argue, more honest) moral evaluations and considerations of responsibility. While others have touched on the implications of these sorts of arguments for different elements of the policy and academic worlds, I would like to focus on their implications for just war theorizing.

Just War Theorizing and a Continuum Approach to Violence

While some just war theorists have engaged seriously with questions of what constitutes violence, what constitutes war, and what constitutes peace (e.g., Walzer 1977; Johnson 1986), many have not and simply assume some constancy to the concepts (e.g., O'Brien 1977; Chesterman 2001). Just war theorizing may not be consistent at defining violence, but it does have several tools for paying attention to the level of violence. Among *jus ad bellum* principles, both proportionality and probability of success pay attention to the level of violence. The *ad bellum* proportionality principle suggests that states are only permitted to make acts of war that are proportional to the severity of the acts that justified the war to begin with (the just cause) (see, e.g., discussions in Walzer 1977; Hurka 2005). The idea is that violence is to be tempered by the evil that inspired it, and that this tempering supposedly limits the level of war by the cause that it is deployed to redress (e.g., Gardam 1993). In theory, the principle of reasonable chance of success also governs the level of violence that is found morally permissible (Childress 1978; Harbour 2011). The principle of reasonable chance of success does not accept the use of any violence, even in the presence of a just cause, if that violence is unlikely to produce a "successful" result (see, e.g., Harbour 2011). Definitions of success vary in the literature, but are largely based on the achievement of justice in one way or another, whether it is redress of the just cause (e.g., Walzer 1977) or the establishment of a just *post bellum* political order (e.g., Williams and Caldwell 2006). Approaches to just war theorizing that emphasize either just peace or *jus contra bellum* (justice against war) also exist, even when they are both less prevalent and less dominant among just war theorists than the traditional categories of *jus ad bellum* and *jus in bello* or the emerging category of *jus post bellum*.[4] The existence (or lack thereof) of a reasonable hope of an ideal end then governs the level (or existence) of permissible violence.

Jus in bello principles also can be understood as trying to moderate the level of violence in war, perhaps more directly than *jus ad bellum* principles. This can be found in the principles regulating during-war proportionality, forbidding *malum in se* (evil in itself) violence, demanding distinction, and requiring military necessity. *In bello* proportionality extends the *ad bellum* proportionality principle, suggesting that a military action not only needs to be proportional in its scale and conceptualization but also in its practice, where the use of violence is supposed to be limited in its scale by its proportion both as a response to the use cause and in reaction to the level of violence that one's opponents use during the conflict (e.g., Walzer 1977; Gardam 1993; Andresen 2014). The principle of forbidding *malum in se* violence is a bar against the use of means considered to be intrinsically heinous, which have included in the literature war rape, torture, cluster-bombing, and so forth (e.g., Skerker 2004; Weigel 2002; O'Brien 1983). In this view, there are some sorts of violence that are never acceptable. Even the use of means considered sometimes acceptable, however, are limited by a number of other *jus in bello* principles. *Jus in bello* principles permit only violence that has military necessity—that which *must* be done (Carnahan 1998; Ramsey 2002). While military necessity is notoriously difficult to determine on both the macro level and the micro level (e.g., Dill and Shue 2012; May 2007), the idea behind the principle is to moderate the level of violence to tailor it to serving the military end toward which it is being deployed. On top of limiting the scope and means of violence and requiring necessity, *jus in bello* principles encourage distinction—asking participants in war to distinguish between combatants and noncombatants, and to provide protection or amnesty to noncombatants (e.g., Walzer 1977; Kalshoven 1981; Primoratz 2007). Academic debates about how deep responsibility for war goes and therefore who is a combatant and who is a civilian (see, e.g., Kinsella 2011; Van Engeland 2011) combine with practical challenges in identifying civilians and combatants during the fighting of wars (e.g., Chesterman 2001; Downes 2008), technical difficulties providing protection to civilians while seeking military ends (e.g., Lightfoot 1994; Goldstoff 2009), and clashes about the level and type of protection that is required to be provided (e.g., Slim 2003; Carpenter 2005) all complicate the distinction principle. Still, in theory, the distinction principle provides the important clarification that, morally, it matters not only what type and level of violence is used but also toward whom the violence is directed.

Jus post bellum also includes some principles moderating the levels and types of violence that just war theorists see as ethically permissible (e.g., Williams and Caldwell 2006; Orend 2000; Stahn 2006). *Jus post bellum* theorizing about the appropriate form of retribution or rehabilitation for aggression being limited to victim compensation and legal redress has the effect of denying legitimation to postwar physical violence (see, e.g., Williams and Caldwell 2006). *Jus post bellum* principles also emphasize measured and reasonable terms for the settlement of war and conflict, minimizing the harm to the party that has lost the war (see, e.g., Bass 2004).

If the multiple dimensions of just war theorizing contain a number of provisions proscribing limits to the forms of violence that should be used, the amount of violence that may be used, and the acceptable targets of violence, why is that not good enough? Just war theorizing cannot itself (or at least cannot definitively) compare particular forms of violence, and it struggles to compare levels and targets of violence. But an understanding that nonviolence is impossible itself also cannot directly address this problem. So the simplest solution to confronting just war theorizing with a continuum approach to violence is to use that understanding of violence as a corrective to the war/not-war dichotomy in just war theorizing and seek a supplementary moral framework to deal with questions of the comparisons of violence along the nonlinear continuum of violence.

This section argues that such an approach is fundamentally both infeasible and undesirable. The understanding of violence laid out in this chapter, I contend, confounds just war theorizing rather than simply complicating it. While it seems on first impression that many of the principles across the subsections of just war theorizing could be stretched to think about violence more broadly, three fundamental problems remain.

First, just war theorizing, even were it expanded to just violence theorizing rather than a near-singular focus on war, fundamentally relies on there being a state of not-war or not-violence as a starting point for analyzing the ethics of embarking on or engaging in war or violence. The *jus contra bellum* and *jus ad bellum* principles of just war theorizing compare starting a war to the constitutive other situation of not-war; *jus post bellum* and just peace assume that war can be brought to an end in a way that produces a situation of peace and nonviolence. In other words, the moral structures of just war theorizing are only capable of comparing levels of violence, types of violence, and targets of violence with an ideal-type of nonviolence as the just warrior's prewar baseline and postwar aspiration. The relative comparisons in just war theorizing are also dependent on the assumption that the "good guy" who started out as innocent of the violence is the one making decisions about engaging in violence, matching the violence of his enemy, and moderating his own violence. The lack of a clean slate beginning or idealizable future necessarily destabilizes just war theorizing's referents for what justice is and how it can be pursued. Replacing the nonviolent or peaceful beginning and end in just war's idealized narratives with a logic interested in minimizing violence does not solve this conundrum because it would not be clear what axes of violence should be minimized and how they might be compared. Expanding the purview of just war to account for a continuum approach to violence expands the impacts of the problem of the lack of a referent for what justice is for just war or just violence, or both. This is because the overlay of multiple interacting forms of violence and sexism complicate the idea of what justice might be, and bring up the ways that legacies of injustice build up to form complicated contexts for moral decisions about and moral judgments of violent actions.

Second, just war theorizing cannot actually easily be expanded to take account of a deeper and wider understanding of violence. Some of the ideas within just war theorizing may have utility for evaluating and moderating a wide variety of acts of violence. For example, the principle of distinction could be applied to regulate the targets of a wide variety of sorts of violence, where understanding who suffers from a particular violent practice, acknowledging it, and targeting violence accordingly is important. The principle of right intent may also be more widely applied, where violent practices may be morally evaluated at least in part by the intent of the person engaging in the violence to redress some other evil. The principle forbidding *malum in se* behavior—intrinsically heinous means—could be expanded to account for violence that is traditionally seen as outside the purview of war (but may well be implicated in the war system). It seems beyond the scope of this chapter to determine what violence is intrinsically heinous, but gender lenses focused on global politics suggest that it might be worth thinking about gender subordination, race subordination, and homophobic violence as intrinsically heinous. Either way, it seems possible to use the *malum in se* principle to evaluate violence across a broad continuum.

If all of these (and potentially other) principles can be applied to a wider understanding of violence, how is just war theorizing inflexible to adopting a broader approach to violence? The answer lies in a more fundamental problem. The just war tradition works in its application to war-proper because of the (gendered) conceptual place of war in people's moral imaginaries. On the one hand, war is hell, and gruesome, and terrible, and combatants, potential combatants, and noncombatants are all aware of that. On the other hand, feminist theorists have extensively documented the ways that states have tied full citizenship to masculinity, and masculinity to military service in defense of the state, its interests, and its innocent women and children (e.g., Elshtain 1987; Sjoberg 2006; Yuval-Davis 1997; McClintock 1993). Feminist theorists (e.g., Enloe 1996; Sjoberg 2013) have also shown that these gendered role expectations have not been immediately transformed with the integration of women into militaries and even into combat roles in those militaries. Just war, and fighting as a just warrior, is not only valorized, it is a key part of the social and political fabric of nationalisms (see, e.g., Elshtain 1987; Stiehm 1982; Sjoberg 2006). This valorization comes in part from claims to justice, but it also relies on an assemblage relationship between (state and individual) masculinities, claims to justice, and practices of war.[5]

The *war* part of just war theorizing, then, is a key substantive element in just war narratizations in several ways. First, as the last paragraph argued, it is not a coincidence that just war theorizing originally and consistently addressed only war rather than violence more generally. The privileging of war is related to the privileged place of war in masculinity, citizenship, and statehood. Second, to the extent that just war narratives have been privileged by states, just war theorizing has served as permissive of war in a variety of contexts (see, e.g., Sjoberg and Peet 2011; Fiala 2008). This permissive function is undesirable, and troubles the translation of just war theorizing's limiting functions. Finally, to the extent that just war theorizing does serve a limiting function, discouraging violent practices, any broad consensus about the need for limiting when war is made and how war is fought can be traced to adherents' concern about the special severity of the violences in war. How bad war is perceived to be is a source of moral consensus for limiting it—no such urgency for ethical consensus may exist for violence popularly understood to be less severe.

These two problems, however, pale in comparison to the last one: the violences of just war theorizing itself. Unmasking the impossibility of just war theorizing's nonviolent beginnings or peaceful ends also shows the problems with the perceived innocence of just war theorizing, just war narratives and discourses, and even just war theorists. Whether deployed in a permissive or limiting way, just war theorizing is itself violent. While I do not have space in this chapter to detail all of the ways in which just way theorizing may be violent, I can provide a couple of examples.

One violence of just war theorizing can be found in its permissive aspects. As I have argued before (Sjoberg 2006), leaders and states use just war discourses to promote and justify acts of war. An example I have used before is former US president George W. Bush's justification of the 2003 invasion of Iraq in a graduation speech at West Point military academy. Bush's (2002) speech contained connotations of the distinction principle (condemning violence against women and children specifically), the principle of just cause (calling the behavior "always and everywhere wrong" [Bush 2002]), the principle of right intent (where "America will call evil by its name" [Bush 2002]), and the principle of last resort (where Bush characterizes it as impossible to reason with Saddam Hussein and therefore characterizes war as the only option). Cues and words from just war theorizing signify to audiences that a war ought to be fought, with little if any regard to the substantive application of just war theories. In this example, and other places, just war discourses can help to justify wars, to rally supporters to wars, and to facilitate the start of wars.

Just war theorizing serves an *in bello* permissive function as well, if unintentionally. As I have written before (Sjoberg and Peet 2011, 2019), the gendered narratives around the distinction principle that frame (presumed female) noncombatants as "beautiful souls" who are understood as innocent both of the evils that started the war and the fighting of the war but nonetheless are essential to the war's occurrence because wars are fought (or justified and characterized as fought) to protect the innocent "beautiful souls" from the enemy (Elshtain 1987). So the distinction principle serves as one of the justifications for fighting (Sjoberg and Peet 2011). But it also has another effect. What feminists have called the "protection racket" (Peterson 1977) is frequently very public—soldiers discuss their motivation to fight in terms of loved ones or the way of life "back home" (Goldstein 2001). "Just warriors" are bound by honor to do this protecting, and the logic of it can be found in most war news and war propaganda. As such, a state's enemies are aware of the value of "beautiful souls" to war efforts and war justifications (Elshtain 1987; Sjoberg 2006). Clausewitz (1989) argues that an opponents' knowledge of belligerents' soft spots, values, or prized possessions matters in those opponents' choices of targets. Clausewitz (1989) calls these things "centers of gravity"—things of high or even ultimate importance to a belligerent regardless of their monetary value. As a strategist, Clausewitz (1989) reports that opponents do (and indeed should) identify belligerents' centers of gravity, then attack them.[6] Empirical evidence suggests that states do exactly that: they victimize civilians intentionally with special attention to women (e.g., Sjoberg and Peet 2011; Harel-Shalev and Daphna-Tekoah 2016). The gendered manifestations of the distinction principle are *in bello* permissive, and constitute a violence of just war theorizing (Sjoberg 2013).

A second (and less complicated) way in which just war theorizing might be seen as violent is in the *ad bellum* principle of reasonable chance of success. One reading of the demand that actors have a reasonable chance of success to engage in acts of war is that it limits violence by stopping futile wars. If the principle has any efficacy, that is likely true. But this reading is limited, because it assumes that violence has been prevented by a party resisting engaging in a futile (if otherwise just) war. Another side of the story of the principle of reasonable chance of success is that it wields war morality discourses to maintain violent and oppressive political configurations in the status quo. If a group (or a person or a state) resists fighting injustice because such a fight would be futile, it legitimizes the injustice and its related violences in practice if not in theory. If we consider the war/not war dichotomy as problematic, the avoidance of a futile war may create a situation of more violence, worse violence, or more poorly targeted violence than fighting such a war would.

The war for the survival and independence of Palestine may be an illustrative example, if we for a second assume that the Palestinian cause meets the other *ad bellum* just war criterion. While many Palestinians fight, looking at how the principle might govern the situation demonstrates the principle's potential violence. It is likely the case that many if not all strategists observing the conflict would argue that Palestine's fight for independence is militarily futile. Whether the tide of national or international opinion, or both, shifts Israel's decisions about occupation is a different question, but in a military fight against the government of Israel, Palestine does not have a reasonable chance to win. So, in theory, using violence in service of the cause of independence would be *de jure* unjust by *jus ad bellum* principles. But it is an open question—one I would never presume to know the answer to—whether engaging in violence in service of the cause of independence brings about a better situation for Palestinians than failing to fight for it. The a priori ruling out of the idea that fighting a futile war may be less violent or less unjust than refraining from fighting such a war presumes a wide variety of value judgments that seem to me impossible for an outsider to make. If Palestinians were actually inspired by the principle of reasonable chance of success to resist fighting, that inspiration itself might be an act of violence on the part of just war theorizing. Even if, situationally, it is not true in this case, it is certainly possible theoretically that there exists a situation of violent oppression less just than the most just futile violence that could be used in the situation. Just war theorizing's ruling that out is itself violent.

Third, and perhaps even simpler, just war theorizing is itself violent because it provides a set of criteria by which violence is understood as not only justifiable but just. Some just war theorists think that the criteria of "just war" do not make violence rise to the level of just, but simply justified or necessary. Still, those words imply moral acceptability. Even if just war theorizing is rigorously applied and treated as strict limitations, it does authorize the practice of violence when/if all the standards are met. Even though that violence might well be "better" violence than violence that does not meet those criteria, it is violence nonetheless.[7]

Fourth, just war theorizing's narrow (and, as discussed above, necessarily so) focus on war combined with the popularity of just war discourses detracts attention from other forms of violence that could be potentially more devastating or more morally problematic, or both, than wars to which just war theorizing might be applied. Feminist and other critical theorizing has pointed out the violences that are contained in constitutive silences and acts of silencing (e.g., Enloe 1996). Other violence that does not receive the same attention as potentially just or potentially unjust wars—the silencing and the silences—are not independent of the dominant just war narratives. They are, instead, two parts of the same whole. The *invisibility* of other violences is a violence of just war theorizing.

Why does it matter that just war theorizing is itself violent? Why does that make it less equipped to serve as a framework for the moral arbitration of the use of violence? The framework of just war theorizing does not have the capacity to recognize (or even recognize and reject) its own violence, or to compare it to other violences that it must analyze.

Because of just war theorizing's reliance on the existence of an alternative state of peace or nonviolence, its inflexibility to accommodate a continuum approach to violence, and its engagement in violence itself, just war theorizing is ill-suited to become a moral framework to analyze violence in global politics once the war/peace and violence/nonviolence dichotomies are deconstructed and discarded. Yet the abandonment of those two (false) dichotomies seems essential for both understanding violence and being realistic about the ethics of violence. Where does that leave the moral analysis of violence in global politics?

Everyday Thinking about the Practices of Violence

I think a question of the magnitude of reinventing literally thousands of years of theorizing war ethics is well beyond the capacity of this chapter, much less its concluding section. Still, faced with evidence that violence operates on a (gendered) continuum that just war theorizing cannot adequately address, I want to propose three preliminary steps for thinking about the ethics of violences in global politics and the study thereof.

I want to suggest that a necessary first step is reflection on the violences involved in one's own participation in research and writing about global politics, as well as one's participation in global politics. Who is excluded from my work? Who is harmed by it? To whom do the beneficiaries of my work do harm? Who is harmed by my planetary footprint? By my purchasing choices? Who benefits? I contend that whatever scholar or whatever citizen of wherever is the subject of these questions, the answer is never that no harm or violence is done. Coming to terms with, trying to understand, trying to manage and shape, and trying to minimize the violence in one's own research and in one's own social and political choices can be a first step toward building an internal moral framework for ourselves.

A second potential step is utilizing feminist principles of dialogue and empathy to look to find moral structures from the bottom up rather than from the top down or from inherited principles (and thereby inherited priorities). Christine Sylvester's (1994) account of empathetic cooperation can teach us to try to understand the moral frameworks of even those who would make the opposite moral choice that we would at almost every turn. Carol Gilligan's (1982) listening guide methodology suggests the importance of hearing nuance in others' needs and choices. Brooke Ackerly's (2000) "multi-sited positionality" demonstrates the utility of getting to know the perspectives of others to broaden one's moral framework and substantive understandings. Laura Shepherd's (2008) deconstruction of security discourses provides a tool for considering the violences in inherited discourses and understanding the complexities of various forms of discursive violence. Annick Wibben's (2010) feminist narrative approach has the capacity to incorporate testimonies, stories, and experiences into dialogues about the ethics of violence. Discussion-based approaches to thinking about moral issues along the continuum of violence are likely to be useful foundational tools.

Perhaps finally (at least for now), an exercise in thinking about and mapping continuum violence—What is violence? What type of violences? What are measures of severity of violence? These exercises have been done in a wide variety of contexts (e.g., Collyer et al. 2007; Marshall 1992; Galtung 1969), and even sometimes have been discussed in the IR/war ethics literatures (e.g., Sarkees and Schafer 2000; Vasquez 2009). That said, an extensive inventory exercise may be useful, especially as it applies to violence in the global political context.

These steps may or may not produce significant insights into a moral framework for evaluating the presence and use of violence in global politics. Whether or not they do, I argue for using a continuum approach to violence to interrogate the war/peace and violence/nonviolence dichotomies, and, following, the capacity and utility of just war theorizing itself.

Notes

1. For examples or descriptions, or both, of emotional violence (Laverne 2014); psychological violence (Lambert 2017); discursive violence (Pratt 2011); epistemic violence (Dotson 2011; Brunner 2018; Teo 2010); violences of exclusion (Choi 2014; Haritaworn, Kuntsman, and Posocco 2013; violences of poverty (Eron, Guerra, and Huesmann 1997); violences of hunger (Scrimshaw 1986); and violences of inclusion (Haritaworn, Kuntsman, and Posocco 2013). This list is not intended to be exhaustive of descriptions of these sorts of violence, and the list of sorts of violence is also not intended to be exhaustive.

2. See, e.g., Moser and Clark (2001) and Gentry and Sjoberg (2015) interrogating the victim/perpetrator dichotomy as a substantive example.

3. For the killing in self-defense example, see, e.g., Quong (2009), McMahan (1994); for the nuclear detonation example, see, e.g., Wasserstrom (1985), Child (1986).

4. For just peace, see, e.g., Chesterman (2001), Allan and Keller (2006); for *jus contra bellum*, see Sharma (2009), Van Steeenberghe (2011).

5. For examples, see, e.g., Belkin (2012), Eichler (2011).

6. Some have argued that the term *center of gravity* should be used only for dual-use targets, but my reading of the concept in Clausewitz disagrees.

7. Feminist responsibility-based approaches to just war theorizing (e.g., Sjoberg 2006; Kellison 2018) have suggested that taking responsibility for violence may be a way to focus on the consequences and severity of violence. While responsibility is one component, acknowledging the violence in even

normative theorizing about violence is in my view an important part of the process of taking responsibility.

References

Ackerly, Brooke. 2000. *Political Theory and Feminist Social Criticism*. Cambridge: Cambridge University Press.

Allan, Pierre, and Alexis Keller. 2006. *What Is Just Peace?* New York: Oxford University Press.

Alnasrawi, Abbas. 2001. "Iraq: Economic Sanctions and Consequences, 1990–2000." *Third World Quarterly* 22(2): 205–18.

Andresen, Joshua. 2014. "Challenging the Perplexity over *Jus in Bello* Proportionality." *European Journal of Legal Studies* 7(1): 18–36.

Arkin, William M. 1999. "The Difference Was in the Details." *Washington Post*, January 17. Accessed August 15, 2018, https://www.washingtonpost.com/wp-srv/inatl/longterm/iraq/analysis.htm?noredirect=on.

Arya, Neil. 2008. "Economic Sanctions: The Kinder, Gentler Alternative?" *Medicine, Conflict, and Survival* 24(1): 25–41.

Bass, Gary J. 2004. "Jus Post Bellum." *Philosophy & Public Affairs* 32(4): 384–412.

Belkin, Aaron. 2012. *Bring Me Men: Military Masculinity and the Benign Facade of American Empire, 1898–2001*. New York: Columbia University Press.

Brock-Utne, Birgit. 1985. *Educating for Peace: A Feminist Perspective*. New York: Pergamon.

Brunner, Claudia. 2018. "Epistemic Violence: Outlining a Term for Peace and Conflict Research." *ZeFKo. German Journal of Peace and Conflict Studies* (SI2): 25–59.

Bush, George W. 2002. "President Bush Delivers Graduation Speech at West Point." June 1, accessed August 25, 2018, https://georgewbush-whitehouse.archives.gov/news/releases/2002/06/20020601-3.html.

Butler, Judith. 1988. "Performative Acts and Gender Constitution: An Essay in Phenomenology and Feminist Theory." *Theatre Journal* 40(4): 519–31.

Butler, Judith. 2004. *Undoing Gender*. New York: Routledge.

Cahill, Lisa S. 2006. *Love Your Enemies: Discipleship, Pacifism, and Just War Theory*. Minneapolis: Fortress Press.

Carnahan, Burrus M. 1998. "Lincoln, Lieber, and the Laws of War: The Origins and Limits of the Principle of Military Necessity." *American Journal of International Law* 92(2): 213–31.

Carpenter, R. Charli. 2005. "Women, Children and Other Vulnerable Groups: Gender, Strategic Frames and the Protection of Civilians as a Transnational Issue." *International Studies Quarterly* 49(2): 295–334.

Chesterman, Simon. 2001. *Just War or Just Peace? Humanitarian Intervention and International Law*. Oxford: Oxford University Press.

Child, James W. 1986. *Nuclear War: The Moral Dimension*. New York: Transaction Publishers.

Childress, James F. 1978. "Just-War Theories: The Bases, Interrelations, Priorities, and Functions of Their Criteria." *Theological Studies* 39(3): 427–45.

Choi, Hyun Jin. 2014. "How Ethnic Exclusion Influences Rebellion and Leader Survival: A Simulation Approach." *Social Science Computer Review* 32(4): 453–73.

Clausewitz, Carl von. (1832) 1989. *On War*. Edited and translated by Michael Howard and Peter Paret. Princeton, NJ: Princeton University Press.

Clegg, Sue. 2013. "The Space of Academia: Privilege, Agency, and the Erasure of Affect." In *Privilege, Agency, and Affect: Understanding the Production and Effects of Action*, edited by Claire Maxwell and Peter Aggleton, 71–87. London: Palgrave.

Cockburn, Cynthia. 2010. "Gender Relations as Causal in Militarization and War: A Feminist Standpoint." *International Feminist Journal of Politics* 12(2): 139–57.

Collyer, Charles E., Frank J. Gallo, Jonathan Cory, Dusty Waters, and Susan Boney-McCoy. 2007. "Typology of Violence Derived from Ratings of Severity and Provocation." *Perceptual and Motor Skills* 104(2): 637–53.

Cuomo, Chris J. 1996. "War Is Not Just an Event: Reflections on the Significance of Everyday Violence." *Hypatia* 11(4): 30–45.

Diehl, Paul F. 2016. "Exploring Peace: Looking beyond War and Negative Peace." *International Studies Quarterly* 60(1): 1–10.

Dill, Janina, and Henry Shue. 2012. "Limiting the Killing in War: Military Necessity and the St. Petersburg Assumption." *Ethics & International Affairs* 26(3): 311–33.

Dotson, Kristie. 2011. "Tracking Epistemic Violence, Tracking Practices of Silencing." *Hypatia* 26(2): 236–57.

Downes, Alexander B. 2008. *Targeting Civilians in War*. Ithaca, NY: Cornell University Press.

Eichler, Maya. 2011. *Militarizing Men: Gender, Conscription, and War in Post-Soviet Russia*. Stanford, CA: Stanford University Press.

Elshtain, Jean Bethke. 1987. *Women and War*. Chicago: University of Chicago Press.

Enloe, Cynthia. 1983. *Does Khaki Become You? The Militarization of Women's Lives*. London: South End Press.

Enloe, Cynthia. 1989. *Bananas, Beaches, and Bases: Making Feminist Sense of International Politics*. Berkeley: University of California Press.

Enloe, Cynthia. 1996. "Margins, Silences, and Bottom Rungs: How to Overcome the Underestimation of Power in the Study of International Relations." In

International Theory, Positivism and Beyond, edited by Steve Smith, Ken Booth, and Marysia Zalewski, 186–202. Cambridge: Cambridge University Press.

Enloe, Cynthia. 2010. *Nimo's War, Emma's War: Making Feminist Sense of the Iraq War*. Berkeley: University of California Press.

Eron, Leonard D., Nancy Guerra, and L. Rowell Huesmann. 1997. "Poverty and Violence." In *Aggression*, edited by Seymour Feshbach and Jolanta Zagrodzka, 139–54. Boston: Springer.

Fazal, Tanisha. 2012. "Why States No Longer Declare War." *Security Studies* 21(4): 557–93.

Fearon, James D., and David D. Laitin. 2003. "Ethnicity, Insurgency, and Civil War." *American Political Science Review* 97(1): 75–90.

Fiala, Andrew. 2008. *The Just War Myth: The Moral Illusions of War*. New York: Rowman & Littlefield.

Frankenberg, Ruth. 1993. *White Women, Race Matters: The Social Construction of Whiteness*. Minneapolis: University of Minnesota Press.

Fussell, Paul. 1975. *The Great War and Modern Memory*. New York: Oxford University Press.

Galtung, Johan. 1969. "Violence, Peace, and Peace Research." *Journal of Peace Research* 6(3): 167–91.

Gardam. Judith G. 1993. "Proportionality and Force in International Law." *American Journal of International Law* 87(3): 391–413.

Gentry, Caron. 2015. "Epistemological Failures: Everyday Terrorism in the West." *Critical Studies on Terrorism* 8(3): 362–82.

Gentry, Caron E., and Laura Sjoberg. 2015. *Mothers, Monsters, Whores: Thinking about Women's Violence in Global Politics*. Rev. ed. London: Zed Books.

Ghobarah, Hazem Adam, Paul Huth, and Bruce Russett. 2003. "Civil Wars Kill and Maim People—Long after the Shooting Stops." *American Political Science Review* 97(2): 189–202.

Gibbons, Chip. 2016. "When Iraq Was Clinton's War." *Jacobin*, May 6, accessed August 15, 2018, https://www.jacobinmag.com/2016/05/war-iraq-bill-clinton-sanctions-desert-fox/.

Gilligan, Carol. 1982. *In a Different Voice*. Cambridge, MA: Harvard University Press.

Goldstein, Joshua. 2001. *War and Gender*. Cambridge: Cambridge University Press.

Goldstoff, Melissa G. 2009. "Security Council Resolution 1820: An Imperfect but Necessary Resolution to Protect Civilians from Rape in War Zones." *Cardozo Journal of Law & Gender* 16(3): 491–518.

Gordon, Joy. 2002. "When Intent Makes All the Difference in the World: Economic Sanctions on Iraq and the Accusation of Genocide." *Yale Human Rights and Development Journal* 5(1): 2–28.

Gordon, Joy. 2010. *Invisible War: The United States and the Iraq Sanctions*. Cambridge, MA: Harvard University Press.

Harbour, Frances V. 2011. "Reasonable Probability of Success as a Moral Criterion in the Western Just War Tradition." *Journal of Military Ethics* 10(3): 230–241.

Harel-Shalev, Ayelet, and Shir Daphna-Tekoah. 2016. "The 'Double-Battle': Women Combatants and Their Embodied Experiences in War Zones." *Critical Studies on Terrorism* 9(2): 312–33.

Haritaworn, Jin, Adi Kuntsman, and Silvia Posocco, eds. 2013. "Murderous Inclusions." Special issue, *International Feminist Journal of Politics* 15(4): 445–452.

Hurka, Thomas. 2005. "Proportionality in the Morality of War." *Philosophy & Public Affairs* 33(1): 34–66.

Johnson, James Turner. 1986. *Can Modern War Be Just?* New Haven, CT: Yale University Press.

Kacowicz, Arie M. 1997. " 'Negative' International Peace and Domestic Conflicts, West Africa, 1957–1996." *Journal of Modern African Studies* 35(4): 367–85.

Kaldor, Mary. 1999. *New and Old Wars: Organized Violence in a Global Era*. Stanford, CA: Stanford University Press.

Kalshoven, Frits. 1981. "Civilian Immunity and the Principle of Distinction." *American University Law Review* 31(3): 855–97.

Kellison, Rosemary. 2018. *Expanding Responsibility for the Just War: A Feminist Critique*. Cambridge: Cambridge University Press.

Kelly, Liz. 1987. "The Continuum of Sexual Violence." In *Women, Violence, and Social Control*, edited by J. Hanmer and M. Maynard. London: Palgrave Macmillan.

Kinsella, Helen M. 2011. *The Image before the Weapon: A Critical History of the Distinction between Combatant and Civilian*. Ithaca, NY: Cornell University Press.

Knight, Ben. 2018. "German Town Stands Up to Neo-Nazis after Swastikas Painted at Scene of Syrian Child's Death." *USA Today*, August 10, accessed August 14, 2018, https://www.usatoday.com/story/news/world/2018/08/10/german-town-schonberg-stands-up-neo-nazis-after-swastika-vandalism/954803002/.

Lacina, Bethany, and Nils Petter Gleditsch. 2005. "Monitoring Trends in Global Combat: A New Dataset of Battle Deaths." *European Journal of Population* 21(2–3): 145–66.

Lake, David A. 2003. "International Relations Theory and Internal Conflict: Insights from the Interstices." *International Studies Review* 5(4): 81–89.

Lambert, Carol A. 2017. "6 Troubling Signs of Psychological Abuse." *Psychology Today*, August 30, accessed August 25, 2018, https://www.psychologytoday.com/us/blog/mind-games/201708/6-troubling-signs-psychological-abuse.

Laverne, Lauren. 2014. "It's Time to Make Emotional Abuse a Crime." *Guardian*, September 7, accessed August 25, 2018, https://www.theguardian.com/lifeandstyle/2014/sep/07/time-to-make-emotional-abuse-a-crime.

Lemon, Jason. 2018. "Unite the Right: Neo-Nazis and Counterprotesters Clash in Washington, D.C., on Charlottesville Anniversary." *Newsweek*, August 12, accessed August 14, 2018, https://www.newsweek.com/hundreds-white-supremacists-rally-second-year-1070003.

Lightfoot, Paul J. 1994. "The Landmine Review Conference: Will the Revised Landmine Protocol Protect Civilians?" *Fordham International Law Journal* 18(4): 1526–65.

Lundahi, Mats. 1989. "Apartheid: Cui Bono?" *World Development* 17(6): 825–37.

Marshall, Linda L. 1992. "Development of the Severity of Violence against Women Scales." *Journal of Family Violence* 7(2): 103–121.

May, Larry. 2007. *War Crimes and Just War.* Cambridge: Cambridge University Press.

McClintock, Anne. 1993. "Family Feuds: Gender, Nationalism, and the Family." *Feminist Review* 44(1): 61–80.

McMahan, Jeff. 1993. "Killing, Letting Die, and Withdrawing Aid." *Ethics* 103(2): 250–79.

McMahan, Jeff. 1994. "Innocence, Self-Defense, and Killing in War." *Journal of Political Philosophy* 2(3): 193–221.

McSorley, Kevin. 2014. "Towards an Embodied Sociology of War." *Sociological Review* 62 (S2): 107–28. https://doi.org/10.1111/1467-954X.12194.

Megret, Frederic. 2011. "War and the Vanishing Battlefield." *Loyola University of Chicago International Law Review* 9(1): 131–58.

Miller, Richard B. 1986. "Christian Pacifism and Just-War Tenets: How Do They Diverge?" *Theological Studies* 47(3): 448–72.

Moreton-Robinson, Aileen. 2011. "The White Man's Burden: Patriarchal White Epistemic Violence and Aboriginal Women's Knowledges within the Academy." *Australian Feminist Studies* 26(70): 413–31.

Moser, Caroline, and Fiona Clark, eds. 2001. *Victims, Perpetrators, or Actors? Gender, Armed Conflict, and Political Violence.* London: Palgrave Macmillan.

Mueller, John, and Karl Mueller. 1999. "Sanctions of Mass Destruction." *Foreign Affairs* 78(3): 43–53.

Muro-Ruiz, Diego. 2002. "The Logic of Violence." *Politics* 22(2): 109–17.

O'Brien, William V. 1981. *The Conduct of Just and Limited War.* New York: Praeger.

O'Brien, William V. 1983. "Just-War Doctrine in a Nuclear Context." *Theological Studies* 44(2): 191–220.

Orend, Brian. 2000. "Jus Post Bellum." *Journal of Social Philosophy* 31(1): 117–37.

Pain, Rachel. 2015. "Intimate War." *Political Geography* 44(10): 64–73.

Peterson, Susan R. 1977. "Coercion and Rape: The State as a Male Protection Racket." In *Feminism and Philosophy*, edited by Mary Vetterling-Braggin, Frederick A. Elliston, and Jane English, 360–71. Totowa, NJ: Littlefield Adams.

Peterson, V. Spike. 2010. "Gendered Identities, Ideologies, and Practices in the Context of War and Militarism." In *Gender, War, and Militarism: Feminist Perspectives*, edited by Laura Sjoberg and Sandra Via, 17–29. Santa Barbara, CA: Praeger Security International.

Pfanner, Toni. 2004. "Military Uniforms and the Law of War." *International Review of the Red Cross* 86(853): 93–130.

Pratt, Mary Louise. 2011. "Violence and Language." *Social Text* Online. May 21, accessed August 25, 2018, https://socialtextjournal.org/periscope_article/violence_and_language_-_mary_louise_pratt/.

Primoratz, Igor, ed. 2007. *Civilian Immunity in War*. New York: Oxford University Press.

Quong, Jonathan. 2009. "Killing in Self-Defense." *Ethics* 119(3): 507–37.

Ramsey, Paul. 2002. *The Just War: Force and Political Responsibility*. Lanham, MD: Rowman & Littlefield.

Reardon, Betty. 1985. *Sexism and the War System*. New York: Teachers' College Press.

Russett, Bruce. 1994. *Grasping the Democratic Peace*. Princeton, NJ: Princeton University Press.

Salbi, Zainab. 2006. *Between Two Worlds: Escape from Tyranny; Growing Up in the Shadow of Saddam*. New York: Penguin.

Sarkees, Meredith Reid, and Phil Schafer. 2000. "The Correlates of War Data on War: An Update to 1997." *Conflict Management and Peace Science* 18(1): 123–44.

Scrimshaw, Nevin S. 1986. "Consequences of Hunger for Individuals and Society." *Federal Proceedings* 45(10): 2421–26.

Shafer, D. Michael. 1994. *Winners and Losers: How Sectors Shape the Developmental Prospects of States*. Ithaca, NY: Cornell University Press.

Shah, Anup. 2002. "Iraq Was Being Bombed during 12 Years of Sanctions." *Global Issues*, April 5, accessed August 15, 2018, http://www.globalissues.org/article/107/iraq-was-being-bombed-during-12-years-of-sanctions.

Sharma, Serena K. 2009. "The Legacy of Jus Contra Bellum: Echoes of Pacifism in Contemporary Just War Thought." *Journal of Military Ethics* 8(3): 217–30.

Sharp, Dustin N. 2012. "Addressing Economic Violence in Times of Transition: Toward a Positive-Peace Paradigm for Transitional Justice." *Fordham International Law Journal* 35(4): 781–815.

Shaw, Anthony, and Ian Westwell. 2000. *The World in Conflict, 1914–1945*. London: Routledge.

Shepherd, Laura. 2008. *Gender, Violence, and Security: Discourse as Practice*. London: Zed Books.

Simons, Geoffrey L. 1999. *Imposing Economic Sanctions: Legal Remedy or Genocidal Tool?* London: Pluto Press.

Sjoberg, Laura. 2006. *Gender, Justice, and the Wars in Iraq*. New York: Lexington Books.

Sjoberg, Laura. 2013. *Gendering Global Conflict: Toward a Feminist Theory of War*. New York: Columbia University Press.

Sjoberg, Laura, and Caron Gentry. 2007. *Mothers, Monsters, Whores: Women's Violence in Global Politics*. London: Zed Books.

Sjoberg, Laura, and Jessica Peet. 2011. "A(nother) Dark Side of the Protection Racket: Targeting Women in Wars." *International Feminist Journal of Politics* 13(2): 163–82.

Skerker, Michael. 2004. "Just War Criteria and the New Face of War: Human Shields, Manufactured Martyrs, and Little Boys with Stones." *Journal of Military Ethics* 3(1): 27–39.

Slim, Hugo. 2003. "Why Protect Civilians? Innocence, Immunity and Enmity in War." *International Affairs* 79(3): 481–501.

Stahn, Carsten. 2006. "'Jus Ad Bellum,' 'Jus in Bello' . . . 'Jus Post Bellum'?—Rethinking the Conception of the Law of Armed Force." *European Journal of International Law* 17(5): 921–43.

Stiehm, Judith Hicks. 1982. "The Protected, the Protector, and Defender." *Women's Studies International Forum* 5(3–4): 367–76.

Strachan, Hew. 2003. *The First World War: Volume 1: To Arms*. Oxford: Oxford University Press.

Strange, Susan. 1994. *States and Markets*. London: Pinter.

Sylvester, Christine. 1994. *Feminist International Relations in a Postmodern Era*. Cambridge: Cambridge University Press.

Teo, Thomas. 2010. "What Is Epistemological Violence in the Social Sciences?" *Social and Personality Psychology Compass* 4(5): 295–303.

Tickner, J. Ann. 1992. *Gender in International Relations*. New York: Columbia University Press.

Tickner, J. Ann. 2001. *Gendering World Politics*. New York: Columbia University Press.

Van Engeland, Anicee. 2011. *Civilian or Combatant? A Challenge for the 21st Century*. New York: Oxford University Press.

Van Steenberghe, Raphael. 2011. "The Law against War or Jus Contra Bellum: A New Terminology for a Conservative View on the Use of Force?" *Leiden Journal of International Law* 24(3): 747–88.

Vasquez, John. 2009. *The War Puzzle Revisited*. Cambridge: Cambridge University Press.

Walzer, Michael. 1977. *Just and Unjust Wars*. New York: Basic Books.

Wasserstrom, Richard. 1985. "War, Nuclear War, and Nuclear Deterrence: Some Conceptual and Moral Issues." *Ethics* 95(3): 424–44.

Weigel, George. 2002. "The Just War Tradition and the World after September 11." *Catholic University Law Review* 51(3): 689–714.

Weinberg, Gerhard L. 1995. *A World at Arms: A Global History of World War II*. Cambridge: Cambridge University Press.

Wibben, Annick T. R. 2010. *Feminist Security Studies: A Narrative Approach*. New York: Routledge.

Williams, Robert E., and Dan Caldwell. 2006. "Jus Post Bellum: Just War Theory and the Principles of Just Peace." *International Studies Perspectives* 7(4): 309–20.

Winter, Jay. 1998. *Sites of Memory, Sites of Mourning: The Great War in European Cultural History*. Cambridge: Cambridge University Press.

Yuval-Davis, Nira. 1997. *Gender and Nation*. London: Sage.

Chapter 3

Contemporary Nuclear Deterrence Dynamics and the Question of Dual Moral Responsibility

THOMAS E. DOYLE II

The question of "locating moral responsibility" is the central theme of this edited volume, and the moral stakes of this inquiry seem particularly high as automated processes or artificial intelligence (AI) are increasingly integrated into states' war-making capabilities.[1] The central question of this chapter is the extent to which contemporary state practices of nuclear deterrence have problematized the moral responsibility of nuclear-armed states. However, the concept of moral responsibility developed in this volume's other chapters seems insufficiently clear in relation to contemporary nuclear deterrence dynamics, which predate the most recent advances in automated weapons systems and AI motivating this volume's inquiries. Thus, this chapter reframes contemporary nuclear deterrence dynamics as a problem of moral agency and moral responsibility.

Assuming a concept of moral responsibility presupposes a concept of moral agency (see, e.g., Eshelman 2014), the chapter's central question becomes: To what extent do contemporary nuclear deterrence dynamics—and the automaticities or AIs integrated into them—entail an inappropriate loss of human moral agency? And if so, is it possible that the relevant AI components of a nuclear deterrent capability have acquired moral agency such that we can describe the deeper problem as one of *dual moral responsibility*? Finally, if moral agency/responsibility is shared between human and artificial agents, and if this loss of human moral agency is inappropriate in some sense, what follows for our considered ethical judgments on the in/appropriateness of nuclear deterrence as a contemporary security strategy?

As I see it, the stakes of this inquiry are significant for those who have contributed to the nuclear ethics literature in International Relations. Since the early days of the Cold War era, commentators have written about nuclear deterrence in terms of moral goods or evils realized (e.g., war prevention vs. massive human and environmental destruction) or moral rights upheld or violated (e.g., national self-defense/the preservation of liberal democracy vs. unjust killing/human rights violations). Many, if not most, commentators largely assume the efficacy of nuclear deterrence as a national or collective security strategy, although critics question this assumption (see, e.g., Barash 2018). Indeed, if nuclear deterrence is not efficacious, then it is morally (and strategically) bankrupt and ought to be abandoned immediately. If it is efficacious, however, then it is crucial to determine the morality of nuclear deterrence and, for this chapter, to determine if it entails an inappropriate loss of *human* moral agency and responsibility to the agencies of AI.

To further specify this chapter's use of human "moral agency" and "moral responsibility," its central question is divided into two narrower ones: Does contemporary nuclear deterrence entail ethical detachment for human actors? And does nuclear deterrence compel human agents to lose the capacities for self-direction and self-governance to nuclear materialities or automaticities, both bureaucratic and technological?

In what follows, I argue that contemporary nuclear deterrence counts as a borderline case of the loss of human moral agency and responsibility. This is to say, it seems to compel a partial delegation of self-direction to "automatic" processes while also retaining a significant measure of human control over the political and military uses of nuclear weapons. Under conditions of deterrence failure—defined as the failure to prevent (nuclear) aggression against vital interests by a deterree—the logic of nuclear reprisal suggests moral nihilism. And this contention concerning moral nihilism itself needs further theorization on the conditions of human moral agency and responsibility if nuclear weapons truly do "explode just war theory" (Walzer 2015, 282). In this chapter I frame the implications of the moral failure of nuclear use as an inappropriate loss of human moral agency as self-governance. Before this argument is advanced, I lay some groundwork by offering a sketch of a conceptual analysis of the concepts of moral agency and moral responsibility. Afterwards, I apply this conceptual analysis to contemporary nuclear deterrence dynamics with a special focus on the United States and the Russian Federation.

Moral Agency and Moral Responsibility: A Conceptual Sketch

Generally, philosophers regard the concept of moral responsibility to entail a concept of moral agency (Eshelman 2014), defined as an actor's capacity to perceive, deliberate, choose, and act on a given choice. Accordingly, a moral agent can anticipate the likely consequences of action that, if produced by a deliberative and decision-making process, could offer reasons for or against taking a proposed course of action. With such capacities, moral agents become morally responsible for their choices, their actions, and the consequences of such actions. To be morally responsible, on the traditional philosophical view, is to be "worthy of a particular kind of reaction—praise, blame, or something akin to these—for having performed it" (Eshelman 2014). Accordingly, one's reactive attitude to another's or to one's own action can be "merit-based" or "consequentialist-based" (Eshelman 2014). Merit-based instances of praise or blame are based on criteria of desert, while consequentialist-based instances are intended to shape the agent's future behavior, that is, to encourage desirable action or dissuade undesirable action. Of course, ascriptions or attributions of praise or blame can count as an expressed reaction that the agent deserves to receive as well as an expression designed to shape the agent's future behavior.

The question now arises: How do we know if an agent deserves praise or blame *or* if it is appropriate to shape their future behavior by an expression of moral reactive attitudes? Most commentators respond by highlighting the agent's role in producing the choices or actions of moral import (Eshelman 2014). To deserve moral praise or blame, an agent must have "pulled the trigger" so to speak, have established the conditions where the "trigger pull" had to occur, or occupy some social office wherein liability for the "trigger pull" can be attributed. Thus, even though it is undoubtedly true that natural and social forces can significantly structure or influence one's deliberations and actions, all valid expressions of moral reactive attitudes must assume some degree of agent-causation (Clarke and Capes 2017). Let us call this element of moral agency "self-direction."

Another way to understand the concept of moral responsibility is that an agent is "responsible for an action or attitude just in case it is connected to her capacity for evaluative judgment in a way that opens her up, in principle, to demands for justification from others" (Eshelman 2014). On this view of moral responsibility, the moral agent must be prepared to offer good reasons for her action to those affected by it. In one sense, we might see attributions or ascriptions of moral responsibility as explicit or implicit calls for justification (Forst 2012). Suppose an agent who "pulled a trigger" suffers condemnation by those negatively affected. Expressions of condemnation are attempts to hold agents morally accountable for their actions, which is an important feature of moral responsibility. In this situation, agents might produce good moral reasons for pulling the trigger. And if such reasons can be produced and accepted, then the justificatory condition of moral responsibility is satisfied. Moreover, our conception of moral agency expands to incorporate an additional element of "self-governance" or Kantian autonomy (Kant 1999; Schneewind 1992). Self-governance is where the self-directed agent offers reasons for action that reveal self-adherence to moral rules and principles. Hence, agency as self-governance both includes and exceeds agency as self-direction.

This sketch of conceptual analysis so far has assumed that "agent" refers to individual human beings. However, the discourse on moral responsibility holds that collective human actors also possess moral agency. This is to say, state officials have the capacity to "bring the state into being" as an "individual" agent with respect to policymaking (Steele 2008, 17–20; Floridi and Sanders 2004, 363). For instance, a state's change from non-nuclear-armed to nuclear-armed status is the self-directed product of internal political and bureaucratic processes such that the state qua agent causes itself to become a nuclear-armed state. Official pronouncements of newly emergent nuclear-armed states often seek to avoid moral censure by invoking exogenous threats as causes of their nuclearization under the rubric of "strategic necessity." Such pronouncements are exercises of shifting the "blame" to state adversaries for their controversial nuclearizing action, even if their decisions were driven by internal economic or normative considerations (see, e.g., Solingen 2007; Doyle 2015a, chap. 2). Even so, once the new nuclear-armed state has publicly justified its controversial choice, a variety of international actors will then express a series of corresponding attributions or ascriptions of moral praise or blame and thereby attempt to hold the given state morally accountable.

Moral Agency as Self-Direction

Does this conceptual analysis of moral agency and responsibility apply to AIs? In their discussion, Floridi and Sanders (2004, 357) apply the concept of agency as self-direction (their word is "autonomy") to the broad field of AIs as the capacity "to change state without direct response to interaction: [to] perform internal transitions to change its state." This definition is important in establishing that AIs, as well as human persons, are self-directing if their internal operations are responsible for changes from an initial to a subsequent state. It would follow, then, that no human or artificial entity is self-directing if its changes of state are the sole products of exogenous forces. Indeed, they contend that "autonomy" or agency requires that an agent (human or artificial) possess "a certain degree of complexity and independence from its environment" (Floridi and Sanders 2004, 357). An agent's change from one state to another could not properly be called "action" if it could be explained exclusively by exogenous forces. Rather, such changes would count only as mere behavior (Clarke and Capes 2017, section 1.3). On their account, then, a semiautomated weapons system's operations are partially autonomous if they are performed independent of human control—e.g., object recognition—and would then count as "action" (Crowder 2018). By contrast, a fully autonomous weapons system would meet each of the conditions of Floridi and Sanders's definition of an autonomous agent. Accordingly, a fully autonomous weapons system

> is a weapon system that could complete an entire targeting engagement cycle by itself. It could go and look for targets, find them, decide on its own, "Yep, this is a valid target that I'm going to engage," and then attack it all by itself without any human involvement whatsoever, or human supervision. (Crowder 2018)

As with the semiautomated weapons system, it would follow that fully automated system operations would count as "action," some of which are presumably of moral import—that is, it can cause moral good or evil and become the object of moral reactive attitudes (Floridi and Sanders 2004, 364).[2]

If this account of moral agency as self-direction is correct, it follows that self-direction is lost if an agent can no longer change their states without direct response to interaction or, similarly, if agent-causation no longer explains such changes. Again, a distinction among kinds of agency forfeiture is important. For instance, military officers directing combat willingly cede agency as self-direction if they put fully automated weapons systems into operation in combat zones while simultaneously retaining formal command.[3] Even so, their ceding of agency to these weapons systems is neither complete nor final, for they might find good reason to withdraw fully or semiautomated systems from combat scenarios if malfunctions put mission objectives at serious risk.[4] On the other hand, an agent might cede agency unwillingly. For instance, an octogenarian or nonagenarian with increasingly poor reflexes and with a history of auto accidents might find their driving privileges removed by the authorities. An elderly person's corresponding loss of agency as self-direction would seem unwilling if, for instance, they signaled significant displeasure toward the authorities. And, unlike the case of the military officer reclaiming direct control of combat operations from semi- or fully automated systems, this elderly person will likely find the loss of agency as self-direction irreversible.

All that is to say that a partial transfer of agency as self-direction to another agent establishes a condition of "dual moral agency" insofar as the control of action is shared between them. A transfer of agency to an AI constitutes its status as a morally responsible agent subject to moral praise or blame. In contemporary contexts of US civilian control of the military and the laws of armed combat, it seems that the outcome of positive moral assessments of AI military action would lead to their retention while negative assessments would lead to their elimination. And, in the latter case, it would seem necessary to infer that the corresponding loss of human moral agency to the relevant AI was inappropriate.

Moral Agency as Self-Governance

What does the foregoing imply for agency as self-governance? Let us recall that Immanuel Kant's conception of autonomy identifies morality with self-governance independent of any conception of self-direction, and in this Kant followed Jean-Jacques Rousseau's contention that moral freedom is "obedience to a law one has prescribed to oneself" (Rousseau 1968, 65; Kant 1999). Indeed, J. B. Schneewind (1998, 3) contends that Kant "invented the conception of morality as autonomy." Hence, the relevance of Kantian moral theory to this chapter's interests over and above consequentialist moral theories (e.g., utilitarianism) or virtue-ethical moral theories (e.g., Aristotle). Kant's conception of autonomy includes the notion of agency as self-direction insofar as moral agents cause their own moral conduct; however, self-governance establishes a higher standard for what counts as truly autonomous action. Before the question of Kantian autonomy can be applied to AIs, we must know more about its conditions of possibility.

According to Kant, one necessary condition of autonomy is *adherence to self-prescribed laws*. Prior to Kant, the traditional Western notion of morality was that of obedience to divine commands as communicated by clerical or political authorities (Schneewind 1998, 3–4). Since both traditional Western and Kantian morality involve obedience to laws (and hence are one part of the concept of governance), their key differences involve the questions of normative origins and moral motivation. For traditional morality, the human self is not a legitimate source of moral obligation since one can exit from any self-imposed duty on the grounds of personal sovereignty. Accordingly, morality requires obedience to non-self-prescribed laws from natural or revealed divine law. In contrast, for Kant, an actor cannot be autonomous if they fail to self-legislate or fail to abide by such self-legislation. As Schneewind (1998, 310) states, "Kant urges each of us to refuse to remain under the tutelage of others. I do not need to rely on 'a book which understands for me, a pastor who has a conscience for me,'" or, we might add, a lawgiver who prescribes courses of action for me.

The preceding paragraph's point can be made more pointedly. On the subject of moral motivation, Schneewind emphasizes that an actor governed by traditional morality is motivated by external incentives or disincentives. Under such conditions, a traditional moralist is not truly self-directing or self-actualizing. In the absence of an external threat of punishment or promises of reward, the traditional moralist would not likely have departed from their "evil" ways. Right action that is the product of fear of punishment does not deserve moral praise as it is not motivated by a good or upright will. By contrast, Kant advanced a rational principle that expressed the moral law and acted as the requisite and immediate motivation (Schneewind 1998, 9–11). As Rainer Forst has argued,

> what is essential to any moral worth of actions [for Kant] is that *the moral law determine the will immediately*. . . . Were the motive for acting morally a heteronomous one (for example,

> an empirical interest), then morality would lose its point, namely, that it is "owed" without qualification. (Forst 2012, 51)

Accordingly, on this first necessary condition of moral agency or autonomy, we find that the source of the laws and the motive for obedience to them must be entirely self-directed.

As suggested earlier, another necessary condition of autonomy or agency as self-governance is adherence to the duty of justification. Kant's inspiration for this element of moral agency is the Rousseauian notion of obedience to self-prescribed laws in which the appropriateness of individual moral action is its consistency with the "general will," which is that collective will of a political community forged by its members' rational assent to a common interest (Rousseau 1968, 49, 64–65). For Rousseau, action contrary to the general will would exhibit arbitrariness or perhaps status privilege or an exclusive factionalism or tribalism. The Kantian categorical imperative has the same justificatory role as Rousseau's general will. Thus, in his analysis of Kantian morality, Forst contends that the demonstration that action is consistent with moral law requires a duty of justification in which each person owes reasons for action to every other person as "ends in themselves" (Forst 2012, 2). From this duty, each person has a corresponding right to justification. The practice of providing reasons for action not only respects the moral status of other persons, it reflects a respect for the "space of norms" from which reasons are given (Forst 2012, 45).

Ontologically, this "space of norms" is composed of regulatory and constitutive norms required for the exercise of moral reactive attitudes that help define the concept of moral responsibility. The list of regulatory norms is quite large and varies according to context, and space constraints prevent this chapter from offering a detailed account of such norms. By contrast, there are two constitutive norms of import on Forst's account. One is *reciprocity*, which requires that no one may refuse the specific demands of others that one demands from others for oneself; otherwise the justification of any action so taken would be arbitrary or anchored on some sort of privilege. The second norm is *generality*, which requires that reasons invoked for basic norms must be shared by and applied equally to all affected persons (Forst 2012, 6–12). Taken together, reciprocity and generality can constrain certain exercises of agency as self-direction, specifically actions taken on pure self-interest. Based on these two constitutive norms, Forst concludes that agency as self-governance is constituted by a self-legislative practice in which the discourses of justification demonstrate adherence to rules instantiating the norms of reciprocity and generality.

One implication of the discussion so far is that the duty of justification necessary for self-governance has a two-part focus—a point that Forst believes Kant's moral theory did not adequately realize (Forst 2012, 53–55). On the one hand, an agent must reason reflexively to determine the suitability of a proposed maxim in relation to the categorical imperative (Forst 2012, 32). Accordingly, each agent must provide to themselves reasons for action of moral import. On the other hand is the aforementioned duty of each agent to provide justification for action to every other affected person as a member of society understood as a "space of reasons" (Forst 2012, 13–14). As Forst argues:

> If we were not individual members of a "space of reasons" in which we must provide each other with justifications capable of withstanding intersubjective normative examination, then we would be like machines that operate within certain allowances laid down by "norms" but could not hold each other accountable for violations of these norms. (Forst 2012, 45)

Forst's reference to machines that "operate within certain allowances" without being able to "hold each other accountable for violations" is a point that will receive greater attention later in the chapter. For now, though, it is important to emphasize how this reference highlights Forst's departure from Kant on the nature of the moral motivation arising immediately from the moral law. Specifically, he contends it is not the "humanity in one's own person of which Kant speaks, [but rather] the only possible ground of moral obligation is the other human being whose humanity demands unconditional respect, simply because he or she is a human being" (Forst 2012, 54). On this view, the duty of justification is a necessary condition of self-governance precisely because "other human beings" bear (1) ultimate moral authority as ends-in-themselves, which motivates (2) their strong right to justification concerning actions that affect them. Indeed, had the ground of moral obligation been located solely in the "humanity in one's own person," then any duty of justification to other human beings would not emerge. In this manner, Forst believes he supplies a missing piece for Kant's conception of morality as autonomy (see also O'Neill 1992, 297).

A third and final necessary condition of agency as self-governance is an agent's moral psychological state, which makes possible self-legislation and the duty and right of justification. Whether it comes by self-discovery or externally sourced moral education, a moral agent exercises an awareness of others as ends-in-themselves. This awareness entails cognition and recognition that another person rightfully "infringes upon one's self-love such that they become visible as a moral person and human being" (Forst 2012, 59–60). Accordingly, a self-governing agent intentionally rejects any privilege of indulging in acts of self-love that are harmful to others, that abandon moral concern for others, or that disregard the norms of reciprocity and generality governing the shared "space of reasons." Only a sentient actor capable of cognizing and recognizing others as ends-in-themselves as the ground of moral reasoning and action is capable of the other two necessary features of autonomy—self-legislation and justification.

On the foregoing account, Kantian autonomy as self-governance is ceded if an agent reverts to a morality of obedience to laws prescribed by others. For instance, one might reenter a relationship of tutelage to religious or political authorities, whose commands replace the Kantian-Forstian method with the nonreflexive and uncritical application of divinely or ideologically based dictates (Schneewind 1992, 310). Such an application of moral formulas exhibits a social automaticity that is contrary to the human freedom conception in the Kantian-Forstian account. Indeed, in this kind of case, agency as self-direction *and* self-governance is abandoned. It is also abandoned if an agent reverts to the practice of acting on heteronomous interests or hypothetical imperatives, which are typical of "self-interest." Heteronomous motives lack an adequate concern for reciprocity and generality necessary for self-governance, as with the egoist who demands something from others that they refuse to grant to others. In the end, an agent's reversion to traditional morality counts as an inappropriate forfeiture of moral self-governance.

It is important to repeat that ceding Kantian autonomy as self-governance does not necessarily mean that agency as self-direction is lost. Indeed, any reversion to self-interested action or the subjection to religious or political moral tutelage requires some exercise of self-direction. Even so, by maintaining self-direction *simpliciter*, agents cede the practice of cognizing and recognizing others as ends-in-themselves who might rightfully infringe on one's self-love. In turn, any duty of justification to others they felt obliged to honor would be based on a heteronomous motive, such as prudence.

What does the foregoing conceptual analysis imply for the chapter's central question of "dual moral responsibility"? First, it does not necessarily imply that a partial or complete loss of self-direction or self-governance involves ethical detachment as defined by ethical neutrality. Agents, for instance, do not necessarily embrace agnosticism or indifference concerning moral value if they abandon self-legislation for the tutelage of clerical or political authority. Instead, they would likely adopt a change of moral perspective on the values of self and others, such as "I am a sinner in need of direction and my clergy are God's envoys." One might note that the phrase "sinner in need of direction" has the grammatical form of a moral reactive attitude that corresponds with moral responsibility. This suggests that an individual human agent who embraces an external moral code and its corresponding moral judgments is "responsible" in the strict causal sense of producing an utterance. However, it does not follow that such expressed judgments are entirely self-directed if, for example, one's embrace of a set of moral reactive attitudes is tightly linked to the external reward and punishment dynamic that can attend religious or ideological indoctrination.

Finally, this conceptual sketch does not yet suggest adequate answers to the relationship between ceding moral agency as self-direction or self-governance and the automaticities of exogenous social or mechanical forces such as AI, even if some have been suggested. Thus, an adequate response to this question must await its application to the question of nuclear deterrence as an instance of security policy, and it is this issue that occupies the attention of the next section.

Nuclear Deterrence and the Question of Ceding Autonomy

Do contemporary nuclear deterrence dynamics compel an inappropriate loss of human moral agency as self-direction or self-governance, especially if they compel a sharing of moral agency among human and artificial agents? To address this issue, we must first summarize some of the salient political and technical features of nuclear deterrence policy, including that of limited nuclear war policy upon which nuclear deterrent threats rely. For the sake of space considerations, the section limits itself to discussing the relevant United States' and Russian programs.

Reviewing Nuclear Deterrence

Despite some changes in United States and Russian nuclear doctrines and the sizes and compositions of their nuclear arsenals over time, the core mission of their nuclear deterrence policies remains largely unchanged since the Cold War.[5] Nuclear-armed states threaten nuclear reprisal strikes to deter their state adversaries from nuclear aggression against their homelands or allies, to deter conventional aggression that might induce nuclear conflict, and, especially for the United States and Russia, to dissuade enemies and even allies from nuclear proliferation, which might increase significantly the odds of nuclear conflict (see, e.g., Defense Intelligence Agency 2017; Gavin 2012). Accordingly, leaders of nuclear-armed states regard these policies as essential means of national (and alliance) security, where "security" is understood as comprising both state survival and independence. Independence (sometimes expressed as "autonomy") is part of this security conception insofar as state leaders seek to be free from foreign domination over their destiny. Thus, at the very minimum, nuclear deterrence aims to provide state leaders possessing this capacity with the greatest degree of self-direction.

It follows then that policies of nuclear deterrence constitute a relationship between deterrer and deterree, where the former holds at risk the latter's governmental and military assets (counterforce deterrence) and even their civilian assets, such as cities and industrial infrastructures (countervalue deterrence). Counterforce deterrence policy has been increasingly understood as conforming to international moral and legal requirements to avoid intentionally targeting or harming innocent civilians, although there are good reasons for thinking that it increases the risks of nuclear war as nuclear-armed deterrees would be afraid to lose their own nuclear capabilities before they could use them in self-defense (Colby 2014, 53). Hence, it is also understood that countervalue nuclear deterrence policy is better tailored for mutual interstate defense and strategic stability (Booth and Wheeler 2008, 140–41).

The success of deterrence policy is dependent on its effectiveness at causing the deterree to refrain from action harmful to the deterrer's security interests. In turn, threat credibility is a combined function of the deterrer's military capabilities and their intention to use them if the deterree "misbehaves." The successful deterrer avoids recourse to armed self-defense, and in the best of circumstances deterrence policy contributes to national and international security in an anarchic international environment. By contrast, the failure of nuclear deterrence compels the deterrer to decide if it will carry out its threats. It is widely believed that nuclear threats must be carried out in cases of deterrence failure, otherwise the deterrer would lose credibility and the ability to deter future nuclear use (see, e.g., Quinlan 2009, 59–67). Additionally, a few have argued that nuclear reprisal strikes are morally required in conditions of catastrophic deterrence failure. Paul Ramsey, for instance, believed that counterforce nuclear war met the *jus in bello* requirements of noncombatant immunity, and thus counterforce nuclear deterrence was morally appropriate (Ramsey 1962). Michael Walzer's conception of the supreme emergency condition, in which a national people's existence faces an imminent threat of extinction, has also become a familiar point of consideration for those inclined to justify nuclear defense and deterrence in moral terms (Walzer 2015, 268–74). Such arguments rely on consequentialist reasons of war termination, some of which also rely on communitarian or liberal reasons concerning the preservation of liberal democratic orders against illiberal aggression (see, e.g., Gauthier 1984; Quinlan 2009, chap. 5; Rawls 1999). For this chapter's interest in determining the relationship between nuclear deterrence and reprisal strategies and the ceding of moral agency to nuclear materialities or automaticities, some additional knowledge about US and Russian procedures for undertaking nuclear retaliation is required.

Nuclear Reprisal and the Loss of Agency as Self-Direction

The question of nuclear defense and deterrence policies with respect to agency as self-direction involves both technological and political considerations. For the United States, the military use of nuclear weapons is entirely dependent upon the command of the president of the United States (see, e.g., Roberts 2016, 12–13, 43; Gardner 2017; Ward 2018). Once the president is informed of an imminent or actual existential attack against the US homeland or its allies covered by extended deterrence guarantees, he opens the "nuclear football" and consults the plans for nuclear reprisal strikes.[6] The president then relays a launch order via the secure phone in the "football" to the US secretary of defense and the commander of the US Strategic Command. The president must also relay the authentication codes located on the credit-card-sized nuclear "biscuit." From there, these orders are communicated to the various launch officers on one or more legs of the US nuclear triad. The time between the president's command and the launch of US nuclear forces could last between five to fifteen minutes, as the system is designed for speedy reaction.

This description of the US process of nuclear reprisal clearly reveals that any concern about loss of human self-direction of US nuclear forces does not arise. This is the case even if some or all launch officers or even the commander of the US Strategic Command refused to obey the president's launch order (Reuters 2017), since agency as self-direction is not obviated merely because intramural conflict occurs among institutional actors. For instance, the president might relieve from duty one or more obstructionist officers and install those who would comply—in which case, the implementation of the president's order is delayed but not overturned. And even if subordinates' disobedience is effective, one may count a failed launch order as a case of institutional self-direction in the sense that Luciano Floridi and J. W. Sanders relate.

For the Russian Federation, the activation of nuclear reprisal strikes after deterrence failure would also likely emerge out of a humanly directed decision by the Russian president (Ward 2018), whereupon the inferences concerning the loss of agency as self-direction would match the US case. On the other hand, Russian nuclear retaliation might also be activated by a semiautomated computer program called Perimeter. This program originated in the early 1980s when Soviet leader Leonid Brezhnev feared US president Ronald Reagan might undertake a nuclear decapitation strike against the Kremlin and deprive it of retaliatory capabilities (Defense Intelligence Agency 2017, 26–27; Bender 2014; Hoffman 2009). Once activated, Perimeter scans to determine if officials in the Russian nuclear command and control system are still active. It is assumed that negative results from Perimeter's scan sufficiently indicates the destruction of the Russian leadership, in which case Perimeter transmits a launch order to its nuclear-armed missile batteries with the first nuclear detonations against enemy targets occurring soon thereafter. It seems that Russian president Vladimir Putin believes in the need of Perimeter for Russian deterrence (Defense Intelligence Agency 2017, 14–19). Additionally, recent news reports suggest that Russia has also developed a prototype for an autonomous nuclear-armed torpedo that, like Perimeter, would provide another automated means of nuclear retaliatory capability if Russian command and control systems were eliminated (Sanger and Broad 2018).

It would thus seem that Perimeter and a fully developed autonomous nuclear-armed torpedo count as a pair of Floridi and Sanders's artificial agents whose actions are morally qualifiable. Even so, it is difficult to draw a simple or straightforward inference concerning the employment of Perimeter or like systems for anything other than a partial loss of human agency as self-direction. On the one hand, any human delegation of nuclear reprisal to Perimeter or similar systems is a voluntary shift from human-self-direction to a nuclear automaticity: that is, state officials have voluntarily ceded immediate directional capabilities to one or more automated weapons systems. On the other hand, Perimeter functions in accordance with the will of its creators. In that sense, Perimeter is roughly analogous to a last will and testament, a legal document in which a deceased person has dictated binding instructions concerning the allocation of their assets and thereby asserts a posthumous capacity of self-direction. Arguably, the legal or bureaucratic automaticity embodied in a last will and testament is a mechanism of self-direction for the human agent. An activation of Perimeter might be considered similarly in relation to state leaders who expect to be killed in the wake of deterrence failure or nuclear reprisal. For these reasons, the case of Perimeter as a nuclear automaticity must count as a borderline case of ceding human moral agency as self-direction.

For US or Russian nuclear deterrence policies to count as model cases of completely ceding human moral agency of self-direction, their national security bureaucracies would have to replace any humanly directed and semiautomatic systems with those controlled by AIs with a capacity for learning and which would not permit any future human direction. Until then, US and Russian leaders will retain personal control over nuclear deterrence and reprisal mechanisms, even if some delegation to semi- or fully automatic nuclear weapons systems occurs. For these reasons, we can reasonably conclude that nuclear deterrence practices, even in the twenty-first century, have not (yet) compelled the loss of human moral agency as self-direction.

However, one might wonder if the US president's direct control of nuclear defense and deterrence is sufficient to demonstrate that agency as self-direction has been achieved, especially if congressional representatives oppose the president's decisions.[7] I take this to be a question about the ontology of the state and the degree to which "liberal state agency" must include some feature of democratic governance (Doyle 2015a). On the one hand, if a constitutional democracy adopts a nuclear deterrence posture in opposition to the majority will of its citizenry, then it is difficult to argue that it counts as an instance of agency as self-direction

(Doyle 2010, 296–97). On the other hand, it seems inappropriate in the context of the current international legal order to say that an authoritarian state that adopts nuclear deterrence is not self-directed, even if we could determine somehow that most its citizens (e.g., North Korea) were opposed to this policy. Indeed, if North Korea succeeds in retaining its nuclear deterrent and therefore its regime in the face of US pressures to the contrary, one might conclude that they have succeeded in establishing a significant expression of political independence—that is, agency as self-direction.

Nuclear Deterrence and the Loss of Agency as Self-Governance

An earlier section of this chapter advanced a general proposition that the loss of agency as self-direction entails the loss of agency as self-governance, although the loss of self-governance does not entail the loss of self-direction. Do contemporary nuclear deterrence dynamics entail the loss of agency as self-governance? Some of the Kantian-inspired moral critiques of nuclear deterrence referenced above provide some answers. The following paragraphs select one of those critiques for detailed consideration, upon which inferences are drawn for the corresponding possibilities of losing self-governance.

In his 1985 article, Steven Lee described nuclear deterrence as a social institution designed to threaten and, if necessary, impose retributive justice for the "crime" of international aggression (Lee 1985).[8] Lee's description makes explicit the relationship between national security and the normative conceptions of justice and law for states engaging in security competition. Lee's critical assessment of nuclear deterrence is anchored on a deontological moral principle that imposes moral side-constraints on institutional action aiming at some collective or common good: the principle of the morality of social institutions (PMSI). It states: "Social institutions are morally justified only if they achieve their social benefit in a way that does not *systematically* violate nonconsequentialist rules, such as those of justice and respect for rights" (Lee 1985, 551; emphasis mine). PMSI thus differentiates between social institutions that occasionally violate justice and respect for rights—but which are nonetheless morally justified in ways Forst's account would accept—*and* those whose rules and practices cannot but routinely violate justice and respect for rights of those affected. Based on this principle, Lee argued that nuclear deterrence systematically violates justice and the respect for rights for those held hostage to nuclear threats, and it does so in three ways.

One is that, ultimately, deterrers cannot avoid *threatening innocent civilians* with nuclear death who are not responsible for the deterree's foreign policy in any significant way but who are nonetheless held as nuclear hostages:

> It is this feature of innocence and not the illegitimacy of the threatener's demands that makes hostage holding wrong. If the tax man threatens your spouse unless you surrender your money, this is just as much a case of hostage holding as if a gunman does the same thing. (Lee 1985, 553)

Lee's distinction between the actor that the deterrer is attempting to control (i.e., the government and military of the deterree) and the actor being threatened (i.e., the deterree's civilian population) recalls debates among just war theorists about the balance between the justice of national self-defense versus the justice owed to innocent parties as expressed by the noncombatant immunity principle (Walzer 2015, 51–73, 138–44; Nye 1986, 99; Ramsey 1962). For Lee, nuclear deterrence policy (and especially countervalue deterrence) categorically dismisses the noncombatant immunity principle and, hence, is a systematic violation of the rights of personal safety owed to innocent third parties by deterrers. Lee notes that nuclear-armed liberal democracies, sensitive to the moral objections against indiscriminate targeting and uses of force, began to shift toward counterforce deterrence postures by developing automated precision-guided delivery systems that could carry lower-yield nuclear devices (Lee 1985, 560–66; see also Lowther 2017; Roberts 2016). Even so, Lee (*contra* Ramsey) contends that counterforce deterrence cannot avoid the systematic violation of the rights of innocent civilians. This is due to the way counterforce nuclear threats induce the deterree to fear a decapitation strike, as related above in the Russian case of Perimeter. Under such a threat, a nuclear-armed deterree is incentivized under crisis conditions to use their nuclear weapons before they lose them. This stance accordingly lowers the threshold of nuclear war initiation and raises significantly the risks of nuclear catastrophe for innocents. The deterrer might insist it does not intend to target innocent civilians, but its insistence is morally indefensible given the deterrer's foreknowledge that large numbers of civilians will be killed by nuclear blast and fire effects.

A second way in which nuclear deterrence violates PMSI is that the act of threatening innocent persons with nuclear death is intrinsically wrong *even if the threat is never carried out*. Some might object that a successful nuclear deterrence policy is conditionally morally justified because it is "bloodless" and therefore harmless (Walzer 2015, 268–84). On this view, the value of realizing national security outweighs the production of fear among innocent foreign civilians, and one might even presume that many citizens will be blissfully unaware of these nuclear threats for most or all of their lives. For Lee, however, the harm of nuclear threats is independent of citizens' knowledge of being nuclear hostages; rather, it is about the increased risk of nuclear death for those who do not deserve it (Lee 1985, 553). Accordingly, the "bloodless deterrence" objection might have more force if people did not experience grave fear or anxiety upon realizing their place in the nuclear crosshairs, but recent events of false alarms of nuclear attack in Hawaii and Japan reveal such fear is felt intensely (Nagourney, Sanger, and Barr 2018).

The third way in which nuclear deterrence violates PMSI is that it raises the risk of nuclear death on innocent citizens *without their consent* (Lee 1985, 554). Lee acknowledges an objection that states do not seek the consent of an enemy's citizenry to fight war against its government or to deter it from taking hostile action. Indeed, prosecuting the just war might not be possible if seeking such consent were necessary. However, for Lee, the indiscriminate nature of nuclear deterrent threats strongly urges the question of citizens' hypothetical consent to a status of nuclear hostages if the promised social benefit was substantial. This, in turn, raises the question of the tacit consent to increased risk of injury or death that citizens accept for all manner of morally justified social institutions, such as the driving of automobiles (Nye 1986, 42–58).[9] For instance, many or all individuals in most countries tacitly accept the risks of driving automobiles as an almost indispensable means of economic and social interaction, strongly suggesting that driving automobiles is morally acceptable under the requirements of PMSI.

However, Lee's concern is not with actors who impose risks on themselves, as in the case of individual operators of automobiles, but with actors who impose increased risks of undeserved punishment on innocent third parties. Accordingly, Lee considers a nonideal amendment to PMSI that would permit an institution that systematically violates nonconsequentialist rules to nonetheless remain morally justified if, in part, there are no alternative institutions or different forms of that institution that could achieve the desired social benefit—that is, national survival and security—while being less unacceptable from a nonconsequentialist viewpoint (Lee 1985, 558). On this amended PMSI, innocent citizens might, even if reluctantly, consent to becoming nuclear hostages of a foreign country if their own national survival and security benefits could not be otherwise obtained. After considering this amended PMSI, Lee returns to his main critique and emphasizes that nuclear deterrers (especially the United States) have alternative, nonnuclear methods of preventing war (Lee 1985, 562–66). Indeed, many recent events strongly suggest that a majority of the world's peoples do not consent to the current regime of great power nuclear deterrence: for example, the humanitarian initiative to ban nuclear weapons, the fear that the lessons of Hiroshima might be fading, and the 2017 Treaty on the Prohibition of Nuclear Weapons (International Committee of the Red Cross 2015; Soble 2016; International Campaign to Abolish Nuclear Weapons [ICAN] 2015; Gladstone 2017). If this latter conjecture is right, then it is not reasonable to claim that innocent people caught in nuclear deterrence's crosshairs are free from injustice and undue risk to their rights.

What does Lee's critique imply for the question of losing agency as self-governance? Again, it is surprisingly difficult to draw simple or straightforward inferences, and one reason for this might be found in differences between Kantian ideal and nonideal moral theory. Specifically, Kantian ideal moral theory strongly suggests that a deterrer employing nuclear threats has abandoned the categorical duty to treat oneself and other persons as ends-in-themselves. Recalling Forst's account, the abandonment of this categorical duty constitutes a voluntary forfeiture of self-governance insofar as it is impossible to simultaneously regard someone else (e.g., a foreign citizen in the deterree's state) as a moral authority who might rightfully impede upon one's self-love and threaten this person with nuclear death. A deterrer also cannot simultaneously regard itself as a moral end and consent (hypothetically) to a regime of nuclear deterrence if better alternatives for national and international security exist. Accordingly, Kantian ideal morality would judge that deterrers have forfeited self-governance altogether by adopting a means of security that keeps innocent people living under a constant threat of annihilation. This would hold true for nuclear democracies as well as nuclear autocracies.

On the other hand, Kantian nonideal moral theory has resources suggesting that, in response to a deterrer's nuclear threat, a non-nuclear-armed deterree may acquire its own nuclear deterrent, even if by doing so it violates international legal obligations.[10] This suggestion rests on an argument by domestic analogy. First, let us assume with Kant that everyone must regard their own person as a moral end as they must regard other persons. Second, let us make use of an argument by analogy that Lee advances in his above-cited article concerning a government tax official's strategy of deterrence to compel a citizen to pay his taxes. In this analogy, the tax man explicitly threatens a citizen in tax arrears; however, his threat raises the risk of harm to the latter's innocent spouse standing nearby. The tax man's threat is clearly disproportionate and violative of the deterree's and spouse's rights on PMSI. In Kantian nonideal terms related to self-governance, the tax man's threats also violate the norms of reciprocity and generality for all involved. This is to say, the tax man would object to being threatened with harm if he himself were in tax arrears. Moreover, he is treating the deterree and the spouse as instruments to obtain the taxes owed, and it is difficult to see how the deterree and the spouse are not pressured to see themselves as mere means. Under these conditions, the deterree and the spouse have a Kantian duty to not permit their rights to be trampled underfoot (Kant 1963, 193–94). The deterree's right of self-defense follows from their self-regarding duty to preserve their rights, as well as any action that might contribute to the restoration of a relationship consistent with reciprocity and generality. A corresponding nonideal reciprocity principle is suggested: *use only those modes of self-defense that remain within the bounds of reciprocity as conditioned by a wrongdoer's offense*. The final element of this analogy is that we must account for two possible outcomes to the deterrence situation: (1) the tax man carries out the threat of harm if the taxes aren't paid or (2) he keeps the deterrent threat active as the deterree pauses to decide among various responses of compliance, direct defiance, or some form of resistance that is less than direct and might not trigger the deterrer's retaliation.

Based on the proposed nonideal reciprocity principle, the deterree is duty-bound to take defensive action if the first outcome occurs, which in extremis might include lethal action. If the second outcome occurs, this principle directs the deterree to not capitulate to the deterrer's demands but instead to advance counterdeterrent threats in kind. The deterree's counterthreat might even be a bluff or lie. Thus, if the tax man threatens to kill the deterree (and the spouse), then the deterree may threaten the same in return. Of course, the deterree is not permitted to issue such counterthreats if the deterrer threatens nonlethal harm.

As I see it, this domestic analogy is applicable to the contemporary practices of nuclear deterrence, especially for cases of great power nuclear deterrence against weaker nuclear-armed or non-nuclear-armed states (e.g., the United States and Iran). If I am correct on this point, then the nonideal reciprocity principle permits the latter kinds of states to adopt tit-for-tat nuclear deterrence postures against the former as much as is practically possible. This is not to say that the deterree is duty-bound on Kantian nonideal theory to acquire nuclear weapons and transform the nuclear deterrence relationship into a mutual one with the deterrer. However, the deterree is duty-bound to undertake a course of action that is consistent with the larger Kantian duty to establish, maintain, or restore the norms of reciprocity and generality necessary for self-governance. Otherwise, the deterree becomes complicit in the deterrer's subversion of its agency as defined by Kantian autonomy (see, e.g., Doyle 2017). In this case, there is no other inference to draw except that uncontested great power nuclear deterrence compels a categorical loss of agency as self-governance for all involved states, democratically governed or not.

A few objections to elements of the foregoing account might be raised.[11] First, one might object that nuclear deterrence lowers the chances of war and the risks of nuclear death, implying that nuclear hostages might willingly consent to any effort to lower such risks. This objection appears based on a strategic assessment of the effects of nuclear deterrence, and hence it suffers from the inherent methodological weaknesses of deterrence theory: that is, trying to prove a negative (why nuclear war has not happened) and the corresponding worry about omitted variable bias (Morgan 2003, 116–71). More importantly, though, the objection misses Lee's point about citizens' consent on becoming nuclear hostages. Thus, even if nuclear deterrence reduces the risk of nuclear war, nuclear threats remain immoral in the same way that threats of murder are immoral even if such threats reduce the chance of murder occurring.

Secondly, one might object that nuclear deterrence can dissuade adversaries in ways that conventional deterrence cannot. This is especially the case if the United States lacks adequate conventional capabilities to punish nuclear aggression. It is difficult to disagree with this objection on its face.[12] Even so, such a conventional capabilities deficit does not constitute a moral argument for nuclear deterrence, nor does it bear upon the question of moral agency as self-governance. The best response to this objection, and one consistent with any imperative to retain agency as self-governance, is that the United States should upgrade its conventional capabilities so that it can respond decisively to nuclear aggression with conventional force.

A third objection relates to the claim above that US nuclear deterrence can be applied to non-nuclear-armed states as well as nuclear-armed states. The objection is that nuclear deterrence is tasked only to keep nuclear-armed states in check. In response, the objection overlooks the degree to which nuclear deterrence has played a central role in nuclear nonproliferation. After all, the US extended nuclear deterrence guarantee not only seeks to dissuade nuclear aggression by states like North Korea or Russia, but it also seeks to dissuade non-nuclear-armed states from acquiring their own nuclear weapons. The foremost case of this has been the US interest in dissuading Germany from acquiring nuclear weapons (Gavin 2012, 57–74). However, even if this third objection is correct, it still doesn't bear upon the moral question of nuclear deterrence as Lee discusses nor on the question of moral agency as self-governance.

A final objection to this application of nonideal Kantian theory to nuclear deterrence relationships is the widespread view that the addition of any new nuclear-weapon state increases significantly the risk of nuclear conflict and, therefore, of mass destruction of larger numbers of innocent people without their consent. Indeed, the objector might insist that a deterree's credible tit-for-tat counter-nuclear-deterrent threat does nothing but add the names of states to the list of those that have acted against moral principle and thus have forfeited moral agency as self-governance. The corresponding question for my view might be: How could a tit-for-tat counter-nuclear-deterrent response (e.g., Iranian nuclearization) count as an action consistent with the (re)constitution of self-governance among states?

On my view, this entirely reasonable objection is best addressed by recognizing that the 1968 Nuclear Non-Proliferation Treaty regime has been transformed by the nuclear-weapon states and their allies from an institution focusing equally on its three core missions (nuclear nonproliferation, the proliferation-free pursuit of nuclear energy, and the commitment to nuclear weapons abolition) into an institution whose focus on nonproliferation has come at the expense of its disarmament mission (Doyle 2017). The Nuclear Non-Proliferation Treaty has become complicit in the nuclear-weapon states' systematic violation of justice and the rights of innocent people worldwide. Considering the nonideal reciprocity principle and the amended PMSI clause, which permits citizens of deterree states to consent (hypothetically) to institutions that systematically violate justice and the rights of persons if there is no alternative institution that can deliver a better social benefit, it can be cogently argued that deterree states that respond in kind to deterrer states necessarily introduce reciprocity into the bilateral relationship and thereby transform both actors into deterrers and deterrees in equal measure. It is only from this political position of relative equality, that is, their mutual hostility or "unsociable sociability," that the conditions for mutual empathy can have a hope of arising (Kant [1795] 1996; Booth and Wheeler 2008, chap. 9). And mutual empathy is necessary for the cognizing and recognizing of the other as a moral end-in-themselves. Otherwise, the sustained power asymmetry that characterizes the contemporary international nuclear order reinforces a bilateral dynamic closer to that of the Athenians and Melians, where the powerful do what they can, the weak do what they must, and the ethics that guides their relationship is one of instrumentalization and moral automaticity.

Conclusion

This chapter has argued that contemporary nuclear deterrence postures count as borderline cases of the loss of human moral agency. It proposes that agency as self-direction is synonymous with the philosophical concept of agent-causation, and therefore nuclear deterrence doesn't compel the loss of state-self-direction of nuclear force. This is even true in contemporary cases of semi- or fully automated nuclear retaliation systems. However, if nuclear-weapon states in the future delegate their nuclear retaliatory systems to fully automated AI systems that they cannot again access and redirect, then they will have irreversibly ceded agency as self-direction. If human beings must retain ultimate control over the use of military force in all its modes, then such a loss of agency is inappropriate.

The chapter also establishes that agency as self-governance is marked by individual or collective actors' determination of their will to choose courses of action consistent with the Kantian categorical imperative to regard self and others as moral ends-in-themselves. Moreover, self-governance is demonstrated by the duty of justification: that is, demonstrating to oneself and others that a course of action satisfies the requirements of reciprocity and generality, which are necessary to show that a subjectively selected course of action simultaneously could in principle win the unforced consent of others in a "space of reasons." An ideal account of self-governance finds that the nuclear deterrer is compelled to cede moral responsibility or human moral agency, since their deterrent action is utterly inconsistent with the categorical imperative. It also follows from the nonideal account of self-governance that an original nuclear deterrer is compelled to surrender some agency as self-governance. Such an outcome cannot but be taken as morally inappropriate. However, the nonideal amendment to PMSI also cuts out a special exception for the deterree whose otherwise impermissible act of counter-nuclear-deterrence is justified since it is intended to (re)establish reciprocity, generality, and a respect for justice and rights that the deterrer originally subverted. For these reasons, deterree counter-nuclear-deterrence counts as a borderline case for the loss of agency as self-governance.

Notes

1. Many thanks to Steven Roach and Amy Eckert for inviting me to contribute this chapter. Many thanks also to the anonymous reviewers whose comments were very helpful. All remaining errors are, of course, my own responsibility.

2. Floridi and Sanders (2004, 364) contend that their account is "neither consequentialist nor intentionalist in nature." However, their definition of morally qualifiable action leans consequentialist in their usage of "good" and "evil" as opposed to the deontological terms of "right" and "wrong" or the virtue-ethical terms of "virtuous" and "vicious."

3. It is important to note that such deployments would not be feasible except in rare cases, for example, missions where autonomous aerial drones must disable enemy radar stations and yet remain immune to efforts to jam signals. See Crowder 2018.

4. One might wonder how the transfer of self-direction from superior officer to those personnel under his or her command compares or contrasts with the delegation of agency to automated weapons systems. This is a subject that deserves a more extensive discussion than what this chapter can undertake; however, it is useful to offer a brief remark here. If a unit commander is put in charge of an operation, she gains some measure of self-direction and her superior officer in a corresponding manner loses some measure of self-direction. However, as a collective actor, there is no loss of human agency as self-direction in such

a transfer of directive operations. This chapter is concerned with the contrast between human and nonhuman exercises of self-direction for the question of "dual" human and nonhuman moral responsibility. Many thanks to an anonymous referee for raising this question.

5. The literature on nuclear deterrence and deterrence theory is too extensive to cite in full here. Recommended works include, but are not limited to, Morgan 2003; Freedman 2004; Schelling 1996; Quinlan 2009; Walker 2012; and George and Smoke 1974. For a well-developed moral defense of nuclear deterrence as a war prevention and international security measure, see Quinlan 2009.

6. During the Cold War, a catastrophic deterrence failure for the United States would have included a Soviet nuclear attack on the US homeland, an invasion of a NATO ally, or (as seen in the 1962 Cuban Missile Crisis) the placement of Soviet nuclear missiles in nearby countries.

7. Many thanks to the editors for raising this question in our correspondence.

8. Lee follows Michael Walzer's framing of "just wars" as those that punish aggression, which is counted as the only "crime" states can commit against each other. See Walzer (2015), especially chapter 2. On how Walzer also saw nuclear deterrence as a vicarious retributive punishment regime, see page 272.

9. It might be that social institutions, such as driving automobiles, require some form of express consent. For instance, the legal requirements of most states in the United States to have a driver's license, and the tasks that prospective drivers must satisfy to get that license, strongly suggests a formula of express consent. For prospective drivers, they consent to obey traffic laws as a condition of receiving a license. They must also consent to purchasing automobile insurance to offset costs to themselves and to others of auto crashes they might suffer.

10. I have developed a more detailed argument for this position in "Reviving Nuclear Ethics: A Renewed Research Agenda for the Twenty-First Century" (Doyle 2010). The revised and condensed account related in the main text is inspired by this article.

11. My thanks to an anonymous reviewer for raising the first three of these objections.

12. Comments by several security policy experts during a recent workshop on possible responses to North Korean nuclear first use at Johns Hopkins University's Applied Physics Laboratory, April 24, 2019. Names are withheld since their remarks were not for attribution.

References

Barash, David P. 2018. "Nuclear Deterrence Is a Myth: And a Lethal One at That." *Guardian*, January 14. Accessed January 15, 2018. https://www.theguardian.com/world/2018/jan/14/nuclear-deterrence-myth-lethal-david-barash.

Bender, Jeremy. 2014. "Russia May Still Have an Automated Nuclear Launch System Aimed across the Northern Hemisphere." *Business Insider Australia*, September 5. Accessed January 11, 2018. https://www.businessinsider.com.au/russias-dead-hand-system-may-still-be-active-2014-9.

Booth, Ken, and Nicholas J. Wheeler. 2008. *The Security Dilemma: Fear, Cooperation, and Trust in World Politics*. New York: Palgrave Macmillan.

Clarke, Randolph, and Justin Capes. 2017. "Incompatibilist (Nondeterministic) Theories of Free Will." *Stanford Encyclopedia of Philosophy*, edited by Edward N. Zalta. Spring. Accessed January 4, 2018. https://plato.stanford.edu/archives/spr2017/entries/incompatibilism-theories/.

Colby, Elbridge A. 2014. "The United States and Discriminate Nuclear Options in the Cold War." In *On Limited Nuclear War in the 21st Century*, edited by Jeffrey A. Larsen and Kerry M. Kartchner, 49–79. Stanford, CA: Stanford University Press.

Crowder, Lucien. 2018. "Don't Fear the Robopocalypse: Autonomous Weapons Expert Paul Scharre." *Bulletin of the Atomic Scientists*, January 10. Accessed January 12, 2018. https://thebulletin.org/don%E2%80%99t-fear-robopocalypse-autonomous-weapons-expert-paul-scharre11423.

Defense Intelligence Agency. 2017. *Russia Military Power: Building a Military to Support Great Power Aspirations*. Accessed January 11, 2018. http://www.dia.mil/Portals/27/Documents/News/Military%20Power%20Publications/Russia%20Military%20Power%20Report%202017.pdf.

Doyle, Thomas E., II. 2010. "Reviving Nuclear Ethics: A Renewed Research Agenda for the Twenty-First Century." *Ethics and International Affairs* 24(3): 287–308.

Doyle, Thomas E., II. 2015a. "When Liberal Peoples Turn into Outlaw States: John Rawls's Law of Peoples and Liberal Nuclearism." *Journal of International Political Theory* 11(2): 257–73.

Doyle, Thomas E., II. 2015b. *The Ethics of Nuclear Weapons Dissemination: Moral Dilemmas of Aspiration, Avoidance, and Prevention*. London: Routledge.

Doyle, Thomas E., II. 2017. "A Moral Argument for the Mass Defection of Non-Nuclear-Weapon States from the Nuclear Nonproliferation Treaty Regime." *Global Governance: A Review of Multilateralism and International Organization* 23(1): 15–26.

Dummett, Michael. 1986. "The Morality of Deterrence." Edited by David Copp. *Canadian Journal of Philosophy* 12: 111–27.

Eshelman, Andrew. 2014. "Moral Responsibility." *Stanford Encyclopedia of Philosophy*, edited by Edward N. Zalta. Accessed April 28, 2019. https://plato.stanford.edu/archives/win2016/entries/moral-responsibility/.

Floridi, Luciano, and J. W. Sanders. 2004. "On the Morality of Artificial Agents." *Minds and Machine* 14: 349–79.

Forst, Rainer. 2012. *The Right to Justification: Elements of a Constructivist Theory of Justice*. Translated by Jeffrey Flynn. New York: Columbia University Press.

Freedman, Lawrence. 2004. *Deterrence*. Cambridge: Polity.

Gardner, Frank. 2017. "Trump and the Nuclear Codes." *BBC News*, January 18. Accessed January 11, 2018. http://www.bbc.com/news/world-us-canada-38651616.

Gauthier, David. 1984. "Deterrence, Maximization, and Rationality." *Ethics* 94(3): 474–95.

Gavin, Francis J. 2012. *Nuclear Statecraft: History and Strategy in America's Atomic Age*. Ithaca, NY: Cornell University Press.

George, Alexander, and Richard Smoke. 1974. *Deterrence in American Foreign Policy*. New York: Columbia University Press.

Gitlin, Todd. 1984. "Time to Move beyond Deterrence." *Nation*, December 22.

Gladstone, Richard. 2017. "A Treaty Is Reached to Ban Nuclear Arms: Now Comes the Hard Part." *New York Times*, July 7. https://www.nytimes.com/2017/07/07/world/americas/united-nations-nuclear-weapons-prohibition-destruction-global-treaty.html.

Hoffman, David E. 2009. *The Dead Hand: The Untold Story of the Cold War Arms Race and Its Dangerous Legacy*. New York: Anchor Books.

International Campaign to Abolish Nuclear Weapons (ICAN). 2015. *Humanitarian Pledge*. December. Accessed July 13, 2016. http://www.icanw.org/pledge/.

International Committee of the Red Cross. 2015. "The Human Cost of Nuclear Weapons." International Review of the Red Cross: Humanitarian Debate: Law, Policy, Action, Geneva. Accessed July 11, 2016. https://www.icrc.org/en/international-review/human-cost-nuclear-weapons.

Kant, Immanuel. 1963. *Lectures on Ethics*. Translated by Louis Infield. New York: Harper & Row.

Kant, Immanuel. 1996. "The Metaphysics of Morals." In *The Cambridge Edition of the Works of Immanuel Kant: Practical Philosophy*, edited by Mary J. Gregor and Allen Wood, translated by Mary J. Gregor, 353–604. Cambridge: Cambridge University Press.

Kant, Immanuel. (1795) 1996. "Towards Perpetual Peace." In *The Cambridge Edition of the Works of Immanuel Kant: Practical Philosophy*, edited by Mary J. Gregor and Allen Wood, translated by Mary J. Gregor, 311–52. Cambridge: Cambridge University Press.

Kavka, Gregory S. 1978. "Some Paradoxes of Deterrence." *Journal of Philosophy* 75(6): 285–302.

Lee, Steven P. 1985. "The Morality of Nuclear Deterrence: Hostage Holding and Consequences." *Ethics* 95(4): 549–66.

Lowther, Adam B. 2017. "The Long-Range Standoff Weapon and the 2017 Nuclear Posture Review." *Strategic Studies Quarterly* 11(3): 34–47.

Machiavelli, Niccolò. 1998. *The Prince*. 2nd ed. Translated by Harvey Mansfield. Chicago: University of Chicago Press.

Morgan, Patrick M. 2003. *Deterrence Now*. Cambridge: Cambridge University Press.

Nagourney, Adam, David E. Sanger, and Johanna Barr. 2018. "Hawaii Panics after Alert about Incoming Missile Is Sent in Error." *New York Times*, January 13. Accessed January 16, 2018. https://www.nytimes.com/2018/01/13/us/hawaii-missile.html?_r=0.

Nye, Joseph S. 1986. *Nuclear Ethics*. New York: Free Press.

O'Neill, Onora. 1992. "Vindicating Reason." In *The Cambridge Companion to Kant*, edited by Paul Guyer, 280–308. Cambridge: Cambridge University Press.

Quinlan, Michael. 2009. *Thinking about Nuclear Weapons: Principles, Problems, Prospects*. London: Oxford University Press.

Ramsey, Paul. 1962. "The Case for Making 'Just War' Possible." In *Nuclear Weapons and the Conflict of Conscience*, edited by John C. Bennett, 143–72. New York: Charles Scribner's Sons.

Rawls, John. 1999. *The Law of Peoples*. Cambridge, MA: Harvard University Press.

Reuters. 2017. "U.S. Nuclear General Says Would Resist 'Illegal' Trump Strike Order." November 18. Accessed January 12, 2018. https://www.reuters.com/article/us-usa-nuclear-commander/u-s-nuclear-general-says-would-resist-illegal-trump-strike-order-idUSKBN1DI0QV.

Roberts, Brad. 2016. *The Case for U.S. Nuclear Weapons in the 21st Century*. Stanford, CA: Stanford University Press.

Rousseau, Jean-Jacques. 1968. *The Social Contract*. Translated by Maurice Cranston. London: Penguin Books.

Sanger, David E., and William J. Broad. 2018. "Pentagon Suggests Countering Devastating Cyberattacks with Nuclear Arms." *New York Times*, January 16. Accessed January 17, 2018. https://www.nytimes.com/2018/01/16/us/politics/pentagon-nuclear-review-cyberattack-trump.html?rref=collection%2Fsectioncollection%2Fpolitics&action=click&contentCollection=politics®ion=rank&module=package&version=highlights&contentPlacement=2&pgtype=se.

Schell, Jonathan. 1982. *The Fate of the Earth*. New York: Alfred A. Knopf.

Schelling, Thomas C. 1966. *Arms and Influence*. New Haven, CT: Yale University Press.

Schneewind, J. B. 1992. "Autonomy, Obligation, and Virtue: An Overview of Kant's Moral Philosophy." In *The Cambridge Companion to Kant*, edited by Paul Guyer, 309–42. Cambridge: Cambridge University Press.

Schneewind, J. B. 1998. *The Invention of Autonomy: A History of Modern Moral Philosophy*. Cambridge: Cambrige University Press.

Soble, Jonathan. 2016. "Fear Sharpens in Japan That Hiroshima's Lessons Are Fading." *New York Times*, May 26. Accessed May 26, 2016. http://nyti.ms/1TYH3XF.

Solingen, Etel. 2007. *Nuclear Logics: Contrasting Paths in East Asia and the Middle East*. Princeton, NJ: Princeton University Press.

Steele, Brent J. 2008. *Ontological Security in International Relations*. London: Routledge.

Walker, William. 2012. *A Perpetual Menace: Nuclear Weapons and International Order*. New York: Routledge.

Walzer, Michael. 2015. *Just and Unjust Wars: A Moral Argument with Historical Illustrations*. 5th ed. New York: Basic Books.

Ward, Alexander. 2018. "Trump Doesn't Have a "Nuclear Button" on His Desk: He Could Easily Attack North Korea Anyway." *Vox*, January 9. Accessed January 11, 2018. https://www.vox.com/world/2018/1/3/16844772/trump-north-korea-button-nuclear-taunt.

Chapter 4

Private Military and Security Companies

Justifying Moral Responsibility through Self-Regulation

SOMMER MITCHELL

Private military and security companies (PMSCs) have remained disadvantaged in their search for legitimacy. Much of this is due to the prohibition of the use of mercenaries and the antimercenary norm, which has evolved through the UN General Assembly and its adoption of more than 100 resolutions criticizing mercenary activities. The Security Council denounced their use in the 1960s and 1970s (Panke and Petersohn 2011, 729). Through the UN, the antimercenary norm, and international treaties, international law developed that made the use of mercenaries illegal, as highlighted by the UN Convention against the Recruitment, Use, Financing, and Training of Mercenaries adopted in 1989 (United Nations General Assembly 1989).

The PMSCs that operated during the 1980s and 1990s were largely illegitimate actors, who routinely violated the antimercenary norm (Petersohn 2014, 3). Sarah Percy (2009, 1) identifies two key components of the antimercenary norm. The first speaks to the morality of mercenaries claiming they are immoral because their actions fall outside the state's legitimate authority to use force. In other words, mercenaries do not have a recognized authority to use force. The second finds them morally problematic because their motivation for fighting wars is driven by selfish, financial gains, which means they fight for themselves rather than for a cause.

However, despite the antimercenary norm and international law, PMSCs were hired regularly by state actors and international organizations to provide support services for military and security operations. The increase in private actors operating in conflict zones skyrocketed in 2003 when the United States invaded Iraq. The steady increase of private actors and their controversial actions in Iraq and Afghanistan caused PMSCs to look for ways to alter the negative perceptions that surrounded them individually and the private security industry (PSI) as a whole. This growing tendency to outsource warfare to PMSCs reveals the (private) struggle to acquire legitimacy via compliance with international antimercenary norms. It is this very struggle, I claim, that shows how PMSCs, and not simply states, have begun to shape and reconstitute the (dispersed) meaning of moral authority in twenty-first-century warfare by altering perceptions of their commitment to human rights protections. My analysis, therefore, will be both empirical and discursive, and focus on the humanizing aspects of the PMSCs struggle for legitimacy and competence in twenty-first-century warfare, as well as how this struggle can be interpreted within the growing tensions between legalist and revisionist just war theory, as discussed in the introduction.

Media reporting, as we shall see, shapes perception of wars (e.g., Iraq and Afghanistan) and incidents involving PMSC contractors. PMSCs realized they had to address this negative perception by seeking moral legitimacy through narrowing the application of the antimercenary norm so that it is no longer associated with PMSCs. No longer being associated with the antimercenary norm is important for PMSCs to distance themselves from the mercenary label and all the limitations that come with it. Gaining legitimacy not only ensures the continuation of the PSI but it also makes the legitimate authority to justifiably use force more likely—a reality that would challenge long-held understandings that only states possess the right to legitimate use of force. However, before PMSCs can lay claim and further complicate the legitimate authority to use force, they must first address their continued association with mercenaries and the negative perception perpetuated by the media.

I claim that PMSCs address this negative perception through the discourse of the Montreux Document, the International Code of Conduct for Private Security Service Providers (ICoC), and the ICoC Association (ICoCA). Drawing on constructivist theory, I shall try to frame the PMSCs' new identity and role in warfare. PMSCs seek legitimacy through the process of initiating, developing, and publicizing the 2008 Montreux

Document, the 2010 International Code of Conduct, and the 2013 ICoC Association. As Anna Leander (2012, 111) states, "Setting standards for PMSCs and making reference to the CoC [Code of Conduct] pertaining to their activities legitimizes their presence both on the ground and in regulatory debates." Their inclusion also furthers their moral legitimacy and provides them with the social license to operate because it normalizes a wide range of actors that practice and speak authoritatively on military matters, which often include security, logistical and weapons support, and training services. Furthermore, PMSCs' involvement in the regulatory debate is beyond doubt because the mainstreaming of ICoC places them at its center (Leander 2012, 111).

Together, the Montreux Document, ICoC, and ICoCA establish a discourse framing the three initiatives as a commitment to understand applicable international law, develop normative standards, and implement oversight and accountability measures regarding PMSC use. However, the three documents provide only an appearance of regulation. In reality, the Montreux Document, ICoC, and ICoCA demonstrate the continuation of a system that benefits the industry and its clients through vague language and self-regulation, thereby furthering the violence that legitimate moral authority and self-defense seeks to monitor and control. In this way, my chapter focuses on the unintended effects of PMSCs' struggle to promote and sustain their moral authority to conduct warfare.

I will therefore first address the PMSCs' response to negative perceptions and controversy by analyzing the process of development and discourse of each initiative. My analysis of media perception will demonstrate how the discourse chosen by media outlets has shaped public understanding of PMSCs. I then present media discourse in newspapers compared to the discourse in the Montreux Document, ICoC, and ICoCA to evidence how PMSCs are seeking moral legitimacy that will enhance perception of their moral responsibility. Furthermore, it illustrates how PMSCs utilize this discourse to reframe the perception of the industry as actors committed to just principles (proportionality).

Establishing the PMSC Discourse

To promote the longevity of the industry, PMSCs had to find a way to alter the public's perception of them as mercenaries and to frame themselves as morally responsible and legitimate security providers.

They thus established their own discursive narrative that addressed many concerns over their regulation and accountability. First, regarding regulation, PMSCs and the PSI altered the discourse they used when characterizing their services. Baum and McGahan (2013) discuss how leaders from Executive Outcomes and Sandline spoke freely about their willingness to engage in combat services. However, this stance was too closely associated with mercenaries and limited their legitimacy both publicly and with clients. By the early 2000s, they stopped referring to services like war and military and started referring to their services as research and intelligence, protective security, and risk management (Baum and McGahan 2013, 22).

Second, PMSCs and the PSI became more vocal about their need for regulation and willingness to be regulated. The rise in public criticism occurred because companies, their employees, and actions taken in the field were frequently in the media.[1] They recognized that, without regulation, the limits to their normative standing, as security providers, would continue (Baum and McGahan 2013).

Both concerns led PMSCs and the PSI to develop regulatory principles and standards. The establishment of their discourse can be divided into three stages. The first is the Montreux Document, developed between 2006 and 2008. The second is the development of the ICoC between 2008 and 2010. Finally, the ICoCA was established from 2011 to the present. The ICoC was not designed to replace the Montreux Document. Instead, the Montreux Document, ICoC, and the ICoCA build on each other in an attempt to provide a comprehensive set of standards for evaluating their moral conduct. To understand fully how the Montreux Document, ICoC, and ICoCA form a narrative that supports the moral legitimization of PMSCs, it is necessary to understand how each initiative came about as well as how each initiative is connected. The process of establishing the discursive narrative unfolds in a relatively linear fashion, starting with the Montreux Document and ending with the ICoCA. However linear the narrative may be, it is still a complex story full of subtleties and ulterior motives, as we shall see.

The Montreux Document

The roots of the ICoC lead back to an initiative started by the Swiss government and the International Committee of the Red Cross (ICRC

2009). A series of meetings, five intergovernmental and four expert, between 2006 and 2008 led to the adoption of the Montreux Document.[2] As of September 2016, the number of participating states had grown to fifty-four (Swiss Federal Department of Foreign Affairs 2014). The Montreux Document draws from a diverse group of sources including the Geneva Conventions, the UN Basic Principles on the Use of Force and Firearms by Law Enforcement Officials, ICRC Study on Customary International Humanitarian Law, existing practices in the industry, and the CoESS/UniEuropa Code of Conduct and Ethics for the Private Security Sector[3] (Cockayne 2009, 402–3). The Montreux Document is directed at states and international organizations, rather than the industry itself, and applies only to armed conflict. Although this document may provide the discourse PMSCs need for seeking legitimacy, a closer look at each section leads to the understanding that the document does little actually to regulate the industry. Nonetheless, the Montreux Document is an important step the PSI and PMSCs needed to take in seeking legitimacy.

Prior to the development of the Montreux Document, PMSCs lived in a legal limbo. No one was taking responsibility for their actions. States did not have laws that applied to contractors working outside the borders of their territory, and since PMSCs were rarely incorporated into the state's armed forces, military law was also not applicable. The Montreux Document served as the first step in regulating the industry, by defining the international legal obligations of potential clients. It did not go far enough to be considered a regulatory agreement, but it was a start in the right direction.

The Montreux Document is divided into three parts, the "Preface," "Pertinent International Legal Obligations Relating to Private Military and Security Companies," and "Good Practices Relating to Private Military and Security Companies." The preface (point 1) lays out the purpose of the document that "certain well-established rules of international law apply to States in their relations with PMSCs." With the very first point, the Montreux Document placed the responsibility for PMSCs and their actions in the field with states. Another purpose of the Montreux Document is to "recall existing legal obligations of States and PMSCs" as well as to "provide States with good practices to promote compliance with international humanitarian law and human rights law during armed conflict" (preface, point 2). The preface also identifies the parties, or potential clients, the document addresses. Important definitions include "contracting states" (those states that directly contract services);

"territorial states" (states where PMSCs actually operate); and "home states" (the states where a PMSC is registered or incorporated). As the remainder of the document suggests, these definitions offer a vital distinction. Until this document, it could be said that the responsibility for PMSCs was in moral flux, passing from state to company or vice versa. By moral flux, I mean that PMSCs seemed to fall loosely outside of any formal set of constraints on their actions/conduct, which could bring them into line with the moral rules of law by which states had long abided. In effect, their legal status as private actors, in placing them beyond the statist scope of the moral codes of warfare, rendered their moral authority problematic. In short, there were few, if any, standards to constrain their actions and to bring their conduct into compliance with the rules of law.

Such moral uncertainty raises the important question of how, in falling outside the legalist and revisionist just war theory, PMSCs' efforts to seek more legitimacy are managing to expose the tensions within just war theorizing, especially as it relates to moral responsibility. By distinguishing between contracting, territorial, and home states, the Montreux Document can and does clearly state the responsibility that each type of state holds regarding PMSCs. The document, for example, stipulates pertinent international legal obligations of each type of state regarding the use of PMSCs (Montreux Document, "Pertinent International Legal Obligations Relating to Private Military and Security Companies," 11–15). Each type of state is addressed in their own subsection of the legal obligations section, yet each point, regardless of subsection, is standard across all types of states. All types of states are obligated not to contract with PMSCs for services that are in opposition to international humanitarian law ("Pertinent International Legal Obligations," points A2 and A3, B9 and B10, and C14 and C15). All states are also obligated to take action such as military regulation or adopting legislation to suppress human rights violations ("Pertinent International Legal Obligations," points A3 and A4, B9 and B10, and C14 and C15). Finally, all states are obligated to investigate, prosecute, extradite, or surrender those suspected of violating international law ("Pertinent International Legal Obligations," points A6, B12, and C17). In appearance, the Montreux Document sets out obligations for all the states that might have interactions with PMSCs. However, all states being obligated to take the same actions regardless of status as contracting, territorial, or home states means it is still possible, maybe even probable, that the actions of PMCSs are not addressed since all parties can say some other party is responsible for regulating violations of international law.

The Montreux Document then addresses good practices for PMSCs and their activities ("Good Practices Relating to Private Military and Security Companies," 16–27). Clarifying good practices is important in terms of regulation and accountability for an industry that often occupies and operates in an unclear area of international law. These obligations and duties are placed with contracting states, territorial states, and home states. All states, regardless of type, are reminded that their responsibilities are nontransferable and that they are obligated to ensure respect for international humanitarian law and human rights and ensure criminal accountability even if that means adopting new legislation. In addition, good practices are described for each type of state to provide guidance and assistance in their relationships with PMSCs both in and outside areas of conflict.

Thus, it is important to note that the Montreux Document does not endorse the use of PMSCs (Leander 2012, 111) nor does it establish new legal obligations (Geneva Academy 2013). It is, however, as Anna Leander (2012, 108) puts it, "the most important interstate initiative pertaining to the governance of the use of force by commercial actors taken" since the UN Mercenary Convention. The document, which was adopted on September 17, 2008 by seventeen states including the United States and the United Kingdom, was a positive first step in moving the PSI toward regulation and accountability. Unfortunately, by focusing on armed conflict, the document leaves a lot of gray space in which PMSCs continue to operate. This gray space includes services that require personnel to be armed, such as guarding locations, providing armed security for individuals, and training police forces. PMSCs have deliberately removed themselves from offering combat services in order to move away from the mercenary label. The Montreux Document only addresses armed conflict, which means that many of the services for which PMSCs are contracted fall outside its scope. Furthermore, the focus on state regulation of the PSI and PMSCs deemphasizes the moral and legal responsibility of individual employees and companies to protect the human rights of individuals, exposing a limitation of the legal tradition of just war theory and the need to revise it to secure a broader and more uniform moral code for other actors and private actors.

International Code of Conduct

For corporations, codes of conduct are voluntary self-imposed obligations that establish normative standards outside the original core business objectives of a company (Rosemann 2008). They are not legally binding, but they often monitor implementation and compliance, which are both subject to a binding procedure. According to Nils Rosemann (2008), voluntarily adopting international standards and submitting to external monitoring procedures complements existing rules and responsibilities. On the other hand, Leander (2012) argues that although there has been a welcome push for socially responsible companies in recent years, focusing on internal standards may undermine the development of binding regulation. Codes of conduct serve as part of a company's public relations, risk management, and sociopolitical contributions. Despite nonbinding regulation concerns, for Rosemann (2008) a code of conduct for PMSCs would obligate companies to comply with human rights and international humanitarian law and provide an implementation and enforcement mechanism. For example, a code of conduct for PMSCs might require a procedure for certification and a company could lose this certification for not implementing or complying with the code of conduct.

Unlike the Montreux Document, which was directed at states, the International Code of Conduct for Private Security Service Providers was established for the PSI itself. Leading up to the development of this code, much of the discourse centered on regulation. Typically, the calls for regulation focused on the need for improved laws to govern and oversee the industry. The purpose of the code, as outlined in the preface (3–4), is to establish "a commonly-agreed set of principles" and affirmed that signatory companies "have a responsibility to respect the human rights of, and fulfill humanitarian responsibilities towards, all those affected by their business activities." The code thus establishes a set of principles and commitments ranging from "responsible provision of security services" to "specific principles regarding the conduct of personnel." It also acknowledges support for the groundwork set in the Montreux Document, committing signatory companies "to the responsible provision of Security Services so as to support the rule of law, respect the human rights of all persons, and protect the interests of their clients" (3). The code states that it is a complementary measure not intended to replace, limit, or alter international or national law (6). Instead, it requires signatory companies, clients, and other stakeholders to create an external independent mechanism charged with governance and oversight of the code through a certification process. This mechanism has grown into the International Code of Conduct Association (ICoCA 2013), which obligated PMSCs to comply with human rights standards and international humanitarian law.

The ICoC was developed over a series of workshops and conferences attended by representatives from PMSCs, industry associations, governments of various countries (including the United States and the United Kingdom), humanitarian organizations, and NGOs. The original ICoC draft was mutually developed by members of the PSI and Swiss Department of Foreign Affairs and facilitated by the Geneva Centre for the Democratic Control of Armed Forces and the Geneva Academy of International Humanitarian Law and Human Rights. The multistakeholder approach to the development of the code suggests a willingness on the part of the PSI to standardize and regulate their services based on the interests of varying parties. As several of these parties are focused on humanitarian law and human rights, this willingness and cooperation reflects the perception that respecting humane concerns, such as humanitarian law and human rights practices, is a high priority for the PSI and PMSCs. This is an important point to make because not being totally driven by market interests frames their image as pursuing humanitarianism/security rather than profits. This lends a degree of credibility and legitimacy to the ICoC.

Initially, the ICoC document answered negative perceptions by providing PMSCs with a way to alter their image by becoming signatory companies. The signature indicated a desire to take responsibility and an endorsement of the document's principles and guidelines that respect human rights and international law. According to the ICoC website, the document, signed on November 9, 2010 by fifty-eight PMSCs, had over seven hundred signatory companies by September 2013.[4] Despite the number of signatures and the appearance of accepting regulations, PMSCs and their industry associations played a large role in the initiation and development of the ICoC. The monopoly role that these PMSCs and their affiliates played in drawing up their rules naturally raises questions regarding its integrity.

In their efforts to be perceived as more legitimate, PMSCs utilized the development of the ICoC to address public concerns[5] (Ralby 2015), many of which were raised in response to specific incidents with high media coverage. The ICoC, for instance, sets standards for specific issues like the use of force, torture, sexual exploitation, and human trafficking. These issues relate to incidents and accusations involving PMSCs still operating today. For example, Blackwater (now Academi) employees were involved in an incident with excessive use of force when they opened fire in Nisoor Square in Baghdad, resulting in seventeen Iraqi civilian deaths in 2007, and CACI and Titan Group employees were part of the 2004 Abu Ghraib prison scandal where prisoners were tortured. This clearly suggests that revising just war theory and its focus on state norms was not simply about protecting individual human rights law but also about adopting a more uniform moral code for all other new actors whose legal and moral status lie beyond the scope of international law.

International Code of Conduct Association

The ICoC Association, an organization that grew out of the implementation measures discussed in the ICoC, governs membership and the certification process. The ICoCA is a multistakeholder initiative that includes a General Assembly, board of directors, and the secretariat, which operates under the executive director. The General Assembly is made up of all members, which include private security companies, governments, and civil society organizations. It serves as a forum for voting on decisions made by the board of directors, such as amendments to the ICoC, requirements for membership, and certification procedures. The assembly also provides a forum for dialogue and discussion related to the ICoC and is tasked with meeting at least once a year. The board of directors includes twelve elected members that represent the interests of all association members. It is the decision-making body of the ICoCA and the executive director is appointed by and executes the board's decisions. The secretariat maintains the records on rules, bylaws, and votes that are necessary for the governance of the ICoCA.[6]

The purpose of the ICoCA, as stated in the Articles of Association, is to "promote, govern, and oversee" the implementation of the ICoC and to promote the responsible provision of security services.[7] The articles include three mandates. The first mandate, certification, gives the ICoCA the responsibility of certifying that a company's policies meet the principles and standards of the ICoC. Certification is an important development for how PMSCs are seeking legitimacy. Certificates, in general, imply something is official or authenticated. Whether the certification process has any real teeth is not the important factor in this case since simply being certified by the ICoCA grants some legitimacy.

At the 2016 annual General Assembly, held on September 29, members agreed that all members will undergo ICoCA certification by September 2018. With the addition of a certification time frame and membership policies, the ICoCA no longer recognizes signatory companies. Now, to be considered in good standing, PMSCs must be members of the association, pay dues, and undergo ICoCA certification. A search on June 24, 2018 discovered ninety-five companies identified as "in good standing" with the association,[8] which is significantly reduced from the seven hundred signatory companies in September 2013. The certification process went into effect on November 1, 2016.[9]

Since the certification process includes a review of company policy and contractor fieldwork, clients are encouraged to require that companies be certified before contracts will be awarded. As more clients include certification requirements in their contracts, the more companies will have no choice but to certify. At the same time, without the cooperation of most clients in the world, the situation with PMSCs that do not certify under the code remains the status quo. For example, out of the thousands of PMSCs operating throughout the world, fewer than one hundred are meeting the requirements for membership in the ICoCA.

The second mandate, reporting, monitoring, and assessing performance, requires that the ICoCA exercise oversight of member companies' performance under the ICoC. This oversight includes external monitoring, gathering information on how companies are operating in the field,[10] and communicating with member companies to address concerns. Under this mandate, member companies are required to provide written assessments of their performance and cooperate in good faith with oversight policies. Mandated oversight of PMSCs supports their efforts in seeking legitimacy because it indicates that companies are constantly monitored and regularly under review. It generates the impression that PMSCs are regulated by and accountable to their governing body.

The third mandate, the complaints process, obligates the ICoCA to maintain a process for handling complaints about member companies' alleged violations of the ICoC. The ICoCA website provides detailed information on who can complain, when a complaint can be made, how to file a complaint, the complaint process, and a timeline for processing complaints.[11] A complaint process also furthers the efforts of PMSCs in seeking legitimacy because it allows victims to voice their grievances. It provides the appearance of accountability for PMSCs.

As with the certification process, the procedures for registering complaints were adopted by members in September 2016. Interestingly, any affected individual or representative can file a complaint. Complainants are informed within thirty days if their claim is accepted for processing. Once it moves forward, the ICoCA assesses if the member companies' grievance mechanism is a viable path, and, if not, determines if there are other fair and accessible grievance mechanisms available to the complainant. Although there is a fleshed-out complaints process, it includes a confidentiality component that requires parties to not disclose matters relating to the allegations and resolutions of the complaint to anyone outside the complaints process. This reality supports the many concerns that the industry's self-regulation with the ICoC and ICoCA is an appearance of regulation and accountability, not an actuality.

For PMSCs, a possible consequence for not meeting its ICoC obligations is loss of legitimate business. Those PMSCs that do not meet the obligations of the ICoC or simply do not certify at all will still have clients seeking their services. Although the process is very new, with a minority of companies having undergone certification, ICoCA discourse is constructing the belief that without certification, the type of client, not to mention the amount of money tied to contracts, will be vastly different. The legitimacy of PMSCs is tied not only to their actions and industry standards but also to the legitimacy of their client base. Accepting contracts with entities that lack legitimacy, such as a dictator or transnational criminal organization, would significantly affect their future pool of clients.

Reframing Media Discourse

Kruck and Spencer (2013, 326) argue that PMSCs care about their image as demonstrated by the hiring of "large public relations firms such as Burson-Marsteller and high-level individual specialists such as Kenneth Starr." Using narrative analysis, Kruck and Spencer (2013) illustrate opportunities and constraints for self-legitimization of security actors. They argue that media narratives indicate how PMSCs are perceived "by a societal opinion elite" that draws on and informs the broader public (Kruck and Spencer 2013, 327). To evidence this claim, Kruck and Spencer analyze PMSCs' self-presentation and media adoption of it. The authors acknowledge an important gap in the literature. Due to limited empirical research on PMSCs successfully establishing a positive image, it is difficult to determine whether PMSCs have discursive power. Conducting a discourse analysis of newspaper media and international agreements helps bridge this gap by highlighting PMSCs' use of international law in conjunction with a code of conduct to seek legitimacy and reframe perceptions.[12]

Brooks and Streng (2012) argue that how the industry is characterized through public discourse affects attitudes toward contractors, toward individuals who work for PMSCs, and how those contractors perceive themselves. Understanding discourse as a concept with concrete categories useful for empirical analysis strengthens the argument presented here. I rigorously analyzed media coverage to establish the discourse used by the media in regard to PMSCs to demonstrate how the PSI addresses negative images and societal concerns through the development of an industry code of conduct.

To find relevant media coverage, I used the Lexis-Nexis Academic database and conducted a newspaper article search for the period between March 20, 2003 (the date the United States invaded Iraq) and November 9, 2010 (the date the ICoC was signed).[13] I chose this time period for three reasons. First, March 20, 2003, is an important date for the United States as it marked the beginning of a long, costly war in Iraq while also fighting a war in Afghanistan, which strained an already taxed military force. Second, this date is also important to PMSCs because it marked an opportunity for massive growth in the PSI. Established companies were awarded numerous contracts that significantly increased their profits. However, these companies struggled to keep up with the demands for so many contracts, which provided the opportunity for many new companies to pop up all over the world. Not only did the private security industry grow in terms of profit, it also increased in terms of size. Finally, November 9, 2010, is important because it is the date the ICoC was signed. Developing a code of conduct provided the PSI a platform to address public concerns raised in newspaper articles. The ICoC serves as the industry's discursive response to media discourse by addressing the most controversial incidents involving PMSCs.

The 188 articles that make up my sample came largely from US and UK newspaper outlets. This is due to many of the larger PMSCs, those with large annual profits, being based in these areas. It is also because these two states award a large number of contracts worth a significant amount of money. Although the sample came from mainstream, large newspaper outlets, there is some variation in their political stance. For example, the *New York Times* tends to be a more liberal publication, where *The Times* of London is more conservative. Prior to my analysis of the articles, the academic literature revealed five major codes: labeling, use of force, regulation, accountability, and torture. Through my analysis of the newspaper articles, the themes that emerged followed these codes. I have chosen to focus my analysis here on the labeling of PMSCs because the process of developing the Montreux Document, the ICoC, and the ICoCA demonstrated PMSCs' willingness to address the regulation and accountability of the industry. Furthermore, this willingness highlights the desire of PMSCs to lose the mercenary label. I determined through my analysis a list of sample terms, listed in table 4.1, that best identified relevant articles.

The Media Label

The media coverage in the years between the US invasion of Iraq (March 20, 2003) and the signing of the International Code of Conduct for Private Security Professionals (November 9, 2010) painted an unsavory picture of PMSCs and their contractors. For example, in 2005, *Frontline* gave "viewers an unprecedented behind-the-scenes look" at KBR, Erinys, and Blackwater in "Private Warriors" (Gaviria and Smith 2005). David Isenberg, writing for the *Huffington Post*, regularly published articles using terms like "Dogs of War" and "Shadow Force," which led to the 2008 publication of his book *Shadow Force: Private Security Contractors in Iraq*. Seymour Hersh published one of the first articles on the torture at Abu Ghraib prison, exposing the role of contractors in the inhumane treat-

Table 4.1. Terms within each theme that identified relevant articles

Labeling
Cowboys
Guns for Hire
Hired Guns
Mercenaries
Mercenary

ment of prisoners in the May 10, 2004 issue of the *New Yorker*. These news reports and others like them raised questions about the number of contractors working in Iraq, human rights violations, cost effectiveness, overcharging, accountability, and transparency.

The public climate for PMSCs and the contractors who work for them was not positive and led to the perception that what PMSCs are hired to do reflects the negative connotations associated with mercenarism like war profiteering, questionable allegiances, and indiscriminate use of force. Furthermore, it led to congressional inquiries resulting in reports such as *Private Security Contractors in Iraq: Background, Legal Status, and Other Issues* (Elsea, Schwartz, and Nakamura 2008) and required that US Central Command produce a quarterly census report on the number of contractors operating in their area of responsibility including Iraq and Afghanistan.[14]

PMSCs' preferred term is *private military* or *private security contractors*, yet they were frequently referred to as mercenaries, hired guns, cowboys, guns for hire, dogs of war, soldiers of fortune, and war profiteers. Newspapers maintained the perception and connection between PMSC contractors and mercenaries with headlines such as "Mercenaries in Trouble Spots to Be Regulated"; "Steroids, Drink, and Paranoia: The Murky World of the Private Contractor"; "Terri Judd on the Guns for Hire Fighting for Business in Iraq and Afghanistan"; "Cowboys Chase Riches in the New Wild West"; and "Blackwater and Its Soldiers of Misfortune." However, some PMSC representatives added to the problem such as when Tim Spicer, the former owner of Sandline International,[15] was quoted as saying "his kind were directly descended from the classic mercenary companies of antiquity" (quoted in Klein 2007, D01).

Table 4.2, which illustrates the number of times authors used negative terminology, shows that the word "mercenaries" was used in 16 percent of the articles while "mercenary" was used in 11 percent.

Table 4.2. Media characterization of PMSCs

Search Term	Number of Instances Used	Number of Articles
Mercenaries	63	30
Mercenary	41	20
Hired Guns	21	13
Cowboys	15	10
Guns for Hire	9	6

Note: This table was compiled using the sample of 188 articles from the Lexis-Nexis Academic search.

As discussed earlier, the label mercenary is perceived as extremely negative and against international law. The reference and characterization of PMSCs as mercenaries was especially problematic for the industry. Though PMSCs were doing their best to lose the mercenary distinction, the media continually referred to them in that way, bringing not only their actions into question but also their legality. For example, in a *Washington Post* article, "Warnings Unheeded on Guards in Iraq: Despite Shootings, Security Companies Expanded Presence," Scott Fainaru (2007) reports that critics "warned that the Pentagon had used an obscure defense acquisition rule to push through a fundamental shift in American war-fighting without fully considering the potential legal and strategic ramifications." Furthermore, the media regularly questioned the use of PMSCs. For example, Norton-Taylor (2006, 16) states:

> The government admits that private security companies are here to stay, and that their operations are likely to increase further as pressures on the armed forces increase. Yet it is keeping the companies at arm's length, apparently concerned about dealing with "mercenaries." The companies, meanwhile, are desperate to shake off what they insist is an outdated and misleading moniker.

Here, he addresses the growing use of PMSCs in Iraq while also putting the number of companies and contractors operating in the country into perspective. For him this illustrates the fine line governments walked regarding PMSCs. Governments, in other words, needed these companies to act as force multipliers. Force multipliers allow military personnel to focus on combat operations while they provide noncombat services. It would have been extremely difficult, if not impossible, for the United States to fight a two-front war in Iraq and Afghanistan without them. Yet however necessary they were to military operations, the US and UK governments kept them on the periphery of their armed forces. This left PMSCs bereft if their employees were captured or killed as well as unregulated with little oversight.

Overcoming the Mercenary Label

As the media label illustrates, the wishes of PMSCs to not be associated with mercenaries were ignored. Academics may have dropped the mercenary label, but the debated categories were not determined by the industry. Furthermore, the media used mercenary frequently to help sensationalize their stories and gain readership. Sensationalized stories about the conduct of contractors overseas tied the negative perception of PMSCs to the mercenary label and contributed to their need for legitimacy. The ICoC and ICoCA provided the perfect opportunity for PMSCs to take ownership of their label.

PMSCs have utilized the ICoC to overcome the mercenary label in three ways. First, they claim ownership of the potential consequences their role as security providers may have. According to the preamble, PMSCs[16] acknowledge that their activities can have "positive or negative consequences for their clients, the local population, the general security environment, and the enjoyment of human rights and rule of law" (ICoC preamble, point 1). This is important to the overall success of the ICoC because PMSCs are often on the defensive about their actions, claiming they are justified or in self-defense. None of that language acknowledges the potentially negative consequences of using PMSCs to fill security needs. This statement inspires a reexamination of an industry that is taking responsibility for their controversial past while moving forward with normative standards that can prevent the kinds of incidents that put the PSI in a negative light to begin with. As this statement appears at the beginning of the document, it serves as a first step in using the ICoC to change discourse.[17]

Second, PMSCs claim their label of choice with the first point in the preamble by referring to themselves as "Private Security Companies and other Private Security Service Providers (collectively PSCs)." The ICoC defines PSCs as "any company whose business activities include the provision of Security Services either on its own behalf or on behalf of another" (ICoC definitions). Security services refer to the "guarding and protection of persons and objects . . . (whether armed or unarmed), or any other activity for which the Personnel of Companies are required to carry or operate a weapon in the performance of their duties" (ICoC definitions). This highlights an important difference in how the Montreux Document and the ICoC label private security providers. Where the ICoC (industry-centered) defines and utilizes the PSC label, the Montreux Document (state-centered) defines and utilizes the PMSC label. For the purposes of the Montreux Document (preface, point 9a), PMSCs are defined as "private business entities that provide military and/or security services irrespective of how they describe themselves." Military and security services include "armed guarding and protection of persons and objects, such as convoys, buildings and other places; maintenance and operation of weapons systems; prisoner detention; and advice to or training of local forces and security personnel" (preface, point 9a).

The difference matters for the PSI for two reasons. First, according to Brooks[18] and Streng (2012, 304) the PMSC label is problematic because it is "inherently faulty and deceptive" since only a small percentage of companies deal with armed security." The authors argue that only 5–10 percent of the industry performs armed security tasks and that the use of "military" is "inaccurate, as the companies are civilian, and it is misleading to imply they have the legal rights and responsibilities of soldiers under international law" (304). Doug Brooks and Hanna Streng both worked for the International Stability Operations Association when they authored "The Stability Operations Industry: The Shared Responsibility of Compliance and Ethics" in 2012. The International Stability Operations Association, the largest international trade association for PMSCs, took part in the development of the Montreux Document and the ICoC. As indicated through their article, they advocated against the all-encompassing label of PMSC. It is in their interest to push a label with a less controversial history as representatives of the leading PMSC trade association. Finally, utilizing PSC instead of PMSC is a good example of the PSI's attempts at changing discourse because one of the most controversial aspects of these companies is their placement in combat. Dropping military from the label, to a degree, diverts attention from the military aspects of the industry.

Lastly, PMSCs utilize the ICoC to overcome the mercenary label by identifying sound principles for the selection and vetting of personnel. The quality of the actual personnel employed by PMSCs was a serious public concern. Unlike issues of the use of force and torture, this concern is not a result of a specific incident. It is a culmination of all the media coverage on the questionable behaviors and hiring practices associated with private contractors. One questionable behavior is individuals applying for positions with PMSCs while falsifying their work experience. In "Clampdown on Rogue Security Guards," McGrory (2004, 1) points out:

> Another concern is that there is evidence of rogue companies having no proper vetting for new applicants. Former British soldiers who signed up to work as bodyguards in Iraq have told *The Times* how they have been sent recruits from Britain

> with no military training and who lied about their backgrounds on application forms.

Another questionable behavior was PMSCs not conducting thorough background checks:

> The private security industry regulator has promised to tighten vetting practices after *The Independent* revealed that the man accused of shooting dead two fellow security contractors in Iraq had a long history of psychiatric illness, was awaiting trial for assault, and had previously been sacked by another private security company. (Judd and Peck 2009, 6)

Finally, participation in the overthrow of governments is questionable behavior that concerned the public as well as policymakers. Hastings (2006, 29) argues that "hired guns" "like Mark Thatcher's merry band who sought to stage a coup in Equatorial Guinea" make headlines "because they have been caught doing ugly and reckless things." All three of these articles illustrate the need for PMSCs and the PSI to create solid practices for the selection and vetting of personnel.

Concern was also raised about the vetting of personnel who are third country nationals. A third country national (TCN) is a person working for a PMSC who is neither a citizen of the hiring state nor of the host state. Table 4.3 illustrates the number of TCNs working in Iraq and Afghanistan from 2008 through 2010. It is easy to see that the number of TCNs increased steadily through this period. More importantly, table 4.3 highlights that there were almost equal numbers of armed and unarmed TCNs. This is significant because the difficulty with hiring TCNs is that criminal background screenings return inaccurate or unavailable records in some countries (Solis 2006). Furthermore, determining the quality, based on training and experience, of TCNs is nearly impossible until they start to operate in theater (Hammes 2011). Industry representatives like Brooks and Streng (2012) argue that only 5–10 percent of contractors are armed. However, the numbers reported in 2009 show nearly as many armed TCNs, 36,438, as unarmed, 39,769, in Iraq. Even if they were the only PMSC contractors armed that year, which is unlikely, that is still a large number of weapons in the hands of individuals who may not have been thoroughly vetted. Interestingly, table 4.3 also illustrates that the US Central Command quarterly census reports stopped differentiating between armed and unarmed contractors after the first two quarters in 2010. This indicates that presenting the number of armed contractors may have been deemed too problematic for reasons such as providing counts on the number of arms in Central Command's area of responsibility or numbers that illustrated there were more armed contractors in Iraq and Afghanistan than policymakers were led to believe and therefore no longer reported.

Singer (2005) argues that although most of the employees of PMSCs are highly qualified individuals, the rush for contractors at the beginning of the war in Iraq led to lax vetting processes, allowing less qualified personnel to slip through the cracks. PMSCs, not the government that awarded the contract, are responsible for training their contractors. It is risky to take unqualified and undertrained employees into a combat zone like Iraq because they could jeopardize missions (Hammes 2011). Furthermore, employees with limited qualifications and training are less predictable in their behaviors and actions. The lack of predictability contributed to the perception that PMSCs are mercenaries, cowboys, and hired guns, which directly influenced their legitimacy as security service providers.

PMSCs understood the basis of this concern and determined that action was necessary. Hurst (2004, 6) reported in "No More Must They Cry Havoc and Let Slip the Dogs of War: Tired of the Trigger-Happy Tag Picked Up in Battle Zones Like Iraq" that

Table 4.3. Private security contractor third country nationals (TCNs) in Iraq and Afghanistan

	Total TCNs Iraq	Armed TCNs Iraq	Total TCNs Afghanistan	Armed TCNs Afghanistan
FY 2008	12,669	12,113	47	20
FY 2009	39,769	36,438	1,333	1302
FY 2010	37,915	16,665*	3,340	1,349*

Note: This table was compiled using Department of Defense USCENTCOM (US Central Command) quarterly census reports for fiscal years 2008, 2009, and 2010. The Department of Defense started reporting the number of private security contractors in Iraq and Afghanistan beginning with the third quarter in 2008.

*Reflects the first two quarters because the Department of Defense stopped separating armed PSCs from total PSCs numbers in September 2010.

> some of the biggest and most powerful British security firms have had enough. They are tired of the allegations and criticisms and have decided to take action. At a conference to be held at Oxford University . . . leading figures in the industry will gather to discuss ways of weeding out the rogue firms in an attempt to create a distinction between the legitimate security companies and the mercenaries.

To help alter the public view that many PMSCs hire indiscriminately, the ICoC includes a section designed to establish standards for selecting and vetting personnel. First, the ICoC states that signatory companies must exercise due diligence in the selection of personnel, regularly assess the continued ability of their personnel to perform their duties, and evaluate the physical and mental fitness of their personnel on a regular basis to ensure they meet appropriate standards (ICoC, "Specific Commitments Regarding Management and Governance," point 45). Second, signatory companies must establish and maintain internal policies and procedures designed to determine the suitability of applicants to carry weapons including checks that they have not

> 1. been convicted of a crime that would indicate that the individual lacks the character and fitness to perform security services pursuant to the principles of this Code;
> 2. been dishonorably discharged;
> 3. had other employment or engagement contracts terminated for documented violations of one or more of the principles contained in this Code; or
> 4. had a history of other conduct that, according to an objectively reasonable standard, brings into question their fitness to carry a weapon. (ICoC, "Specific Commitments Regarding Management and Governance," point 48)

For clarity, the ICoC also states that "disqualifying crimes may include, but are not limited to, battery, murder, arson, fraud, rape, sexual abuse, organized crime, bribery, corruption, perjury, torture, kidnapping, drug trafficking or trafficking in persons" (ICoC, "Specific Commitments Regarding Management and Governance," point 49). Third, signatory companies will require that all applicants and employees authorize access to prior employment and government records as a condition of employment and engagement. Finally, all the principles outlined in the ICoC for the selection and vetting of personnel also apply to the selection and vetting of subcontractors (ICoC, "Specific Commitments Regarding Management and Governance," point 51).

The "Selection and Vetting of Personnel" section of the ICoC includes more standards than the "Use of Force" and the "Prohibition of Torture" sections combined, indicating that the stakeholders responsible for developing the ICoC considered setting standards for the selection and vetting of personnel important. Many of the concerns with private contractors may be eliminated altogether if the selection and vetting process can weed out individuals with questionable past behaviors and those who lack the appropriate training and experience.

Conclusion

PMSCs have worked hard at constructing an alternative narrative through discourse that disengaged them from the mercenary label. The Montreux Document, ICoC, and ICoCA helped PMSCs establish their desired narrative by demonstrating their commitment to address public concerns over the questionable behavior of their contractors and the selection and vetting of personnel. I claimed that their struggle for legitimacy showed how tensions for revising just war theory in terms of uniform moral code could be evidenced by the expanding/evolving code of conduct for PMSCs over the past several years.

Having thus analyzed the development and discourse of each initiative and presented an empirical analysis of media coverage that supported the argument that PMSCs are seeking legitimacy through the discourse of the ICoC, I elaborated on the discursive aspects of the just war principle of legitimate authority by showing how PMSCs have sought to foster a perception of competence and humane concern. More importantly, I illustrated the importance of PMSCs' need to own their label so their claim that they are not mercenaries is no longer undermined by media coverage and negative perceptions. Finally, I concluded that PMSCs' struggles for moral legitimacy and competence in twenty-first-century warfare illustrates movement toward their ultimate goal of legitimate authority.

Notes

1. Public criticism is discussed fully in later sections.

2. Full title: *The Montreux Document on Pertinent International Legal Obligations and Good Practices for States Related to the Operations of Private Military and Security Companies during Armed Conflict*. The document can be found in English, French, Spanish, or Chinese at https://www.icrc.org/en/publication/0996-montreux-document-private-military-and-security-companies.

3. CoESS is the industry association for PMSCs in the European Union and the Code of Conduct can be found at http://www.coess.org/_Uploads/dbsAttachedFiles/Code_of_Conduct_and_Ethics_EN.pdf.

4. The association has moved beyond signatures and now has a membership process. This means those signatory companies that have not applied for membership are no longer fulfilling their commitments.

5. For the purposes of this argument, public includes civilians as well as policy makers.

6. For more information on the ICoCA's General Assembly, executive director, and secretariat, see http://www.icoca.ch/en/icoc-association and http://www.icoca.ch/en/articles_of_association #article-9-executive-director-and-secretariat.

7. https://www.icoca.ch/en/articles_of_association#article-2-purpose.

8. http://icoca.ch/en/membership?private_security_companies=companies&op=Search&view_type=list&form_id=_search_for_members_filter_form.

9. http://icoca.ch/en/news/2016-AGA.

10. The articles mandate that the secretariat is responsible for gathering information from available sources including the public.

11. https://www.icoca.ch/en/complaints.

12. Further research beyond the scope of this work is needed to fully determine what success PMSCs' attempts at improving their image and seeking legitimacy have had.

13. Using the search terms "private security," "private military," "Iraq," and "Afghanistan," I collected articles from those newspapers that returned twenty-five or more articles: *Guardian*, *Washington Post*, *Times* (London), *Independent*, and *New York Times*. This resulted in a sample of 188 articles.

14. US Central Command, *Contractor Support of U.S. Operations in the USCENTCOM Area of Responsibility to Include Iraq and Afghanistan*, http://www.acq.osd.mil/log/PS/CENTCOM_reports.html.

15. Tim Spicer is the founder of Aegis Defense Services. More importantly, he is the former owner of Sandline International, which is the now defunct firm that was tied to delivering weapons during Sierra Leone's civil war known as the Arms to Africa scandal.

16. In the first part of the preamble, the industry identifies private security companies and private security service providers collectively as PSCs.

17. The Geneva Academy of International Humanitarian Law and Human Rights published an Academy Briefing titled *The International Code of Conduct for Private Security Providers*. In this briefing, the writers examine each paragraph of the ICoC, providing insight into its language and legal applicability.

18. Brooks is the founder and president emeritus of the International Stability Operations Association, which is the largest international trade association for PMSCs.

References

Baum, J. A. C., and A. M. McGahan. 2013. "The Reorganization of Legitimate Violence: The Contested Terrain of the Private Military and Security Industry during the Post–Cold War Era." *Research in Organizational Behavior* 33: 3–37.

Brooks, D., and H. Streng. 2012. "The Stability Operations Industry: The Shared Responsibility of Compliance and Ethics." *Criminal Justice Ethics* 31(3): 302–18.

Clifford, S. 2008. "Success Overcomes Qualms in E.A.'s Iraq-Themed Game." *New York Times* [online], April 21. Lexis-Nexis Academic search, November 23, 2016.

Cockayne, J. 2009. "Regulating Private Military and Security Companies: The Content, Negotiation, Weaknesses, and Promise of the Montreux Document." *Journal of Conflict Security & Law* 13(3): 401–28.

Elsea, J. K., M. Schwartz, and K. H. Nakamura. 2008. *Private Security Contractors in Iraq: Background, Legal Status, and Other Issues*. Congressional Research Service Report for Congress, RL32419, August 25.

Fainaru, S. 2007. "Warnings Unheeded on Guards in Iraq: Despite Shootings, Security Companies Expanded Presence." *Washington Post* [online], December 24, 2nd ed., A section. Lexis-Nexis Academic search, November 23, 2016.

Gaviria, M., and M. Smith. 2005. "Private Warriors." *Frontline*, PBS.

Geneva Academy of International Humanitarian Law and Human Rights. 2010. *International Code of Conduct for Private Security Service Providers*, November 9. https://www.icoca.ch/sites/all/themes/icoca/assets/icoc_english3.pdf.

Geneva Academy of International Humanitarian Law and Human Rights. 2013. *Academy Briefing no. 4: The International Code of Conduct for Private Security Service Providers*. August. http://www.geneva-academy.ch/docs/publications/briefing4_web_final.pdf.

Hammes, T. X. 2011. "Private Contractors in Conflict Zones: The Good, the Bad, and the Strategic Impact: DTIC Document." https://apps.dtic.mil/dtic/tr/fulltext/u2/a536906.pdf.

Hastings, M. 2009. "We Must Fight Our Instinctive Distaste for Mercenaries: The Iraq Bubble Has Burst but the Need for Private Security Companies Will Not Go Away." *Guardian* [online], August 2. Lexis-Nexis Academic search, November 23, 2016.

Hersh, S. 2004. "Torture at Abu Ghraib." *New Yorker* 80(11) (May 10): 42–53.

Hurst, C. 2004. "No More Must They Cry Havoc and Let Slip the Dogs of War: Tired of the Trigger-Happy Tag Picked Up in Battle Zones Like Iraq." *Independent* [online], November 28. Lexis-Nexis Academic search, November 23, 2016.

International Code of Conduct Association. 2013. *Articles of Association*. September 20. http://www.icoca.ch/en/articles_of_association.

International Code of Conduct Association. 2014. *About the ICoC*. http://www.icoc-psp.org/About_ICoC.html.

International Committee of the Red Cross. 2009. *The Montreux Document on Pertinent International Legal Obligations and Good Practices for States Related to Operations of Private Military and Security Companies during Armed Conflict*. August. https://www.icrc.org/en/publication/0996-montreux-document-private-military-and-security-companies.

Isenberg, D. 2009. *Shadow Force: Private Security Contractors in Iraq*. Westport, CT: Praeger Security International.

Judd, T. and T. Peck. 2009. "Security Industry to Review Vetting after Report on Murder Suspect: Case of Daniel Fitzsimons Highlights Need for Change." *Independent* [online], August 15. Lexis-Nexis Academic search, November 23, 2016.

Kinsey, C. 2006. *Corporate Soldiers and International Security: The Rise of Private Military Companies*. New York: Routledge.

Klein, A. 2007. "For Security in Iraq, a Turn to British Know-How: With U.S. Contract up for Grabs, Congresswoman Requests Audit of Major Bidder." *Washington Post* [online], August 24. Lexis-Nexis Academic search, November 23, 2016.

Kruck, A., and A. Spencer. 2013. "Contested Stories of Commercial Security: Self- and Media Narratives of Private Military and Security Companies." *Critical Studies on Security* 1(3): 326–46.

Leander, A. 2012. "What Do Codes of Conduct Do? Hybrid Constitutionalization and Militarization in Military Markets." *Global Constitutionalism* 1(1): 91–119.

McGrory, D. 2004. "Clampdown on Rogue Security Guards." *Times* [online], August 30. Lexis-Nexis Academic search, Novembeer 23, 2016.

Norton-Taylor, R. 2006. "Fears over Huge Growth in Iraq's Unregulated Private Armies: Mercenaries 'Outnumber UK Soldiers Three to One'; Security Companies Are Unaccountable, Say Critics." *Guardian* [online], October 31. Lexis-Nexis Academic search, November 23, 2016.

Panke, D., and U. Petersohn. 2011. "Why International Norms Disappear Sometimes." *European Journal of International Relations* 18(4): 719–42.

Percy, S. 2009. "Private Security Companies and Civil Wars." *Civil Wars* 11(1): 57–74.

Petersohn, U. 2014. "Reframing the Anti-Mercenary Norm: Private Military and Security Companies and Mercenarism." *International Journal* (August 21): 1–19. https://doi.org/10.1177/0020702014544915.

Ralby, I. 2015. "Accountability for Armed Contractors." *Fletcher Security Review* 2(1): 1–7.

Rosemann, N. 2008. "Code of Conduct: Tool for Self-Regulation for Private Military and Security Companies." Geneva Centre for the Democratic Control of Armed Forces. http://dcaf.ch/Publications/Code-of-Conduct-Tool-for-Self-Regulation-for-Private-Military-and-Security-Companies.

Sandholtz, W. 2008. "Dynamics of International Norm Change: Rules against Wartime Plunder." *European Journal of International Relations* 14(1): 101–31.

Singer, P. W. 2005. "Outsourcing War." *Foreign Affairs* 84: 119–32.

Solis, W. 2006. *Rebuilding Iraq: Actions Still Needed to Improve the Use of Private Security Providers*. GAO-06-865T. Washington, DC: US Government Accountability Office. http://www.gao.gov/products/GAO-06-865T.

Suchman, M. C. 1995. "Managing Legitimacy: Strategic and Institutional Approaches." *Academy of Management Review* 20(3): 571–610.

Swiss Federal Department of Foreign Affairs. 2014. *Participating States of the Montreux Document*. November 12. https://www.eda.admin.ch/eda/en/fdfa/foreign-policy/international-law/international-humanitarian-law/private-military-security-companies/participating-states.html.

United Nations General Assembly. 1989. "International Convention against the Recruitment, Use, Financing and Training of Mercenaries. A/RES/43/34, 72nd Plenary Meeting, December 4, UN, New York." www.un.org/documents/ga/res/44/a44r034.htm.

US Central Command (USCENTCOM). 2008–10. "Contractor Support of U.S. Operations in the USCENTCOM Area of Responsibility to Include Iraq and Afghanistan." http://www.acq.osd.mil/log/PS/CENTCOM_reports.html.

PART II

AUTONOMOUS WEAPONS SYSTEMS AND MORAL RESPONSIBILITY

Chapter 5

The Rights of (Killer) Robots

DAVID J. GUNKEL

The list of moral and legal issues regarding the development and potential deployment of lethal autonomous weapons systems (LAWS) is as complex as it is interesting. "How will the use of robot weapons," as Rob Sparrow (2007, 66) asks, "affect the ways in which wars are fought, the level and nature of casualties, and the threshold of conflict? What sort of decisions should they be allowed to make? How should they be programmed to make them? Should we grant non-human intelligent agents control of powerful weapons at all? If a system is intelligent enough to be trusted with substantial decision making responsibility in battle, should it also be granted moral standing? Should they then be granted rights under the Geneva Conventions?"

The vast majority of existing research takes up and addresses the kinds of questions with which Sparrow begins—questions regarding safety, liability, and responsibility (e.g., Asaro 2012; Johnson and Axinn 2013; Krishnan 2013; Lokhorst and van den Hoven 2012; Noorman and Johnson 2014; Sharkey 2012). When questions of moral standing and rights are in play, these are typically limited to inquiries regarding the status and rights (or, more often than not, the potential for the violation of these rights) of human combatants and noncombatants—those individuals and communities that are on the receiving end of military action undertaken with and by such mechanisms. Very little (almost nothing, in fact) has been published on the last two questions that appear in Sparrow's litany—questions concerning and inquiring about the moral status and even the rights of battlefield robots. In fact, this phrase "the rights of battlefield robots" already sounds wrong and somewhat discordant with contemporary efforts and initiatives regarding international conflict and the development and deployment of what have been called (mobilizing a science fiction–inspired sense of drama) "killer robots."

It is the purpose and objective of this chapter to consider and to make a case for this other, apparently marginal set of concerns by responding to a seemingly simple and direct question: Can or should killer robots have rights? This question, however, is not just any question. At the beginning—even before beginning—we should be clear about the inherent difficulty of even articulating such a query. As David Levy (2005, 393) has pointed out, "the notion of robots having rights is unthinkable," presumably because (1) it is something that is unable to be thought, insofar as the very concept of robots having rights strains against common sense or good scientific reasoning; or (2) it is to be purposefully avoided as something that must not be thought, insofar as it is a kind of prohibited idea or blasphemy that would open a Pandora's box of problems and therefore should be suppressed or repressed (to use the common psychoanalytic terminology). Whatever the reason(s), there is something of a deliberate decision and concerted effort not to think—or at least not to take as a serious matter for thinking—the question of robot rights. But there are good reasons for considering the rights of robots in general and the rights of battlefield robots in particular.

Standard Operating Presumptions

In confronting and contending with other entities—whether other human persons, animals, the natural environment, or technological artifacts—one inevitably needs to distinguish between those beings *who* are in fact moral/legal subjects and *what* remains a mere thing or object. As Jacques Derrida (2005, 80) explains, the difference between these two small and seemingly insignificant words—"who" and "what"—makes a big difference, precisely because it parses the world of entities into two camps: those Others who can and should have a legitimate right to privileges, claims, powers, and immunities[1] and mere things that are and remain objects, instruments, or artifacts to be used without further consideration.

Though every moral and legal system institutes and operationalizes distinctions between *who* is a subject of legitimate consideration and *what* is not—or between what are, in existing legal terminology, "persons" and "property"—a good example/illustration of this decision-making and its consequences is available in the rules governing international armed conflict. International humanitarian law (IHL) comprises a set of rules that seeks to control and limit the impact and effects of war. IHL is based on international treaties like the four Geneva Conventions and their Additional Protocols and a substantial body of "customary law" that is binding on all states and parties engaged in hostile actions. IHL is currently administered, catalogued, and overseen by the International Committee of the Red Cross (ICRC), which was named the controlling authority in the fourth Geneva Conventions of 1949.

Animals

IHL in both name and substance is exclusively anthropocentric (Nowrot 2015; de Hemptinne 2017). Indicative of this exclusivity is what the IHL stipulates (or more importantly, does not stipulate) regarding nonhuman animals. "Despite the fact that certain animals are quite frequently allowed or required to 'participate directly in hostilities' in the sense of Article 43 (2) of the 1977 Geneva Convention Protocol I, they are not granted the rights and do not have the obligations deriving from the legal status of combatants under international humanitarian law" (Nowrot 2015, 136). Animals have accompanied human beings into battle since ancient times. The chronicles and tales of human conflict include a menagerie of other living beings: camels, horses, donkeys, elephants, dolphins, sea lions, birds, and dogs. These "animals soldiers" (a moniker that is itself the site of significant conflict over the presumed role and status of the animal) have fought alongside human beings and have occupied a number of important positions on the battlefield. "Canine assistants," as Julie Carpenter (2015, 43) recounts, "have been used in defense for transport, weapons detection, communications, and as comfort to soldiers on and off the battlefield. As of 2012, there were a reported 2,700 dogs serving with the US military worldwide, with 600 of those active in designated war zones." Despite this service, however, animals are virtually absent from the documents that comprise IHL. As Jérôme de Hemptinne (2017, 272) critically pointed out, "being deeply anthropocentric, international humanitarian law (IHL) largely ignores the protection of animals."

Exceptions to this are informative, mainly because they are those kinds of exceptions that prove the rule. In the existing documents, animals are explicitly addressed in just two places.

- Article 18, first and third paragraphs, of the 1949 Geneva Convention III indicates that prisoners of war retain possession of "all effects and articles of personal use, except arms, horses, military equipment and military documents" (ICRC 2018, Rule 49). In this context, one specific species of animal, horses, are formally listed as "military equipment" that can be legitimately seized from captured human combatants and disposed of as "war booty."

- "The only explicit reference to animals in general is," as Nowrot (2015, 136) points out, "hidden in Article 7 (1) lit j of the 1996 Protocol II on Prohibitions or Restrictions on the Use of Mines, Booby-Traps and Other Devices to the Convention on Prohibitions or Restrictions on the Use of Certain Conventional Weapons Which May be Deemed to be Excessively Injurious or to Have Indiscriminate Effects of 10 October 1980, prohibiting the use of booby-traps and other devices which are in any way attached to or associated with 'animals or their carcasses.' "

Additionally "article 53 of Geneva Convention IV," as de Hemptinne (2017, 276) argues, "prohibits the destruction by the occupying power of private and public properties, except in cases of absolute military necessity. This provision could provide minimum protection to certain animals when considered to be items of private or public property." The extension of the article to animals, and civilian livestock in particular, is codified in Article 54 of the subsequent 1977 Geneva Convention Protocol I addressing the "protection of objects indispensable to the survival of the civilian population and, in this regard, prohibiting inter alia the destruction and removal of 'livestock.' " This protection, however, is enacted not for the sake of the animal but for the purpose of respecting the property rights of its human owner(s). It is, therefore (and at best), a form of indirect protection for a human noncombatant who has the right to possession and use of a particular item of personal property.

Other Things

Animals, as both Nowrot (2015) and de Hemptinne (2017) demonstrate, have been and remain largely excluded from the protections codified in

IHL. The situation is (or would seem to be) worse for robots and other autonomous machines. As the other of the animal other (Gunkel 2012), machines are also excluded or marginalized from the IHL protections granted to both combatants and noncombatants. Under the current rules, things like battlefield robots assisting soldiers in the field are considered "military equipment" that may be seized and destroyed without impunity: "The parties to the conflict may seize military equipment belonging to an adverse party as war booty" (ICRC 2018, Rule 49). Under this stipulation, "military equipment" has been interpreted to designate "all movable State property captured on the battlefield" including "arms and ammunition, depots of merchandise, machines, instruments and even cash" (ICRC 2018, Rule 49).

There are some notable exceptions. But these exceptions, once again, only serve to prove the rule. Some objects, specifically "objects used for humanitarian relief operations," are protected. But these protections are granted not for the sake of the object but for the intended human recipients of the relief this object supports or makes possible. As summarized in ICRC (2018) Rule 32:

> State practice establishes this rule as a norm of customary international law applicable in both international and non-international armed conflicts. This rule is a corollary of the prohibition of starvation (see Rule 53), which is applicable in both international and non-international armed conflicts, because the safety and security of humanitarian relief objects are an indispensable condition for the delivery of humanitarian relief to civilian populations in need threatened with starvation. In that framework, this rule is also a corollary of the prohibition on deliberately impeding the delivery of humanitarian relief (see commentary to Rule 55), because any attack on, destruction or pillage of relief objects inherently amounts to an impediment of humanitarian relief.

Other objects, like "articles of personal use" and "protective gear" belonging to and in the possession of captured combatants and other persons hors de combat cannot be seized and shall remain in their possession (ICRC 2018, Rule 49).

Finally, there are various articles, protocols, and guidelines that stipulate protections for the natural environment and significant cultural objects and sites. But these stipulations and restrictions are instituted not for the sake of the object per se but for the human populations that depend on and value these objects.

> With regard to these and other relevant provisions, it needs to be emphasized in the present context, that the regime of international humanitarian law does not in general protect the environment per se but is first and foremost intended to avoid negative consequences for the affected (human) civilian population indirectly caused by damage to the environment. This is vividly illustrated by the additional requirement as enshrined in Article 55 (1) of the 1977 GC [Geneva Convention] Protocol I stipulating that damage to the natural environment is only of relevance for the purposes of this provision if it is intended or may be expected also to "prejudice the health or survival of the population." (Nowrot 2015, 134)

The same is true of cultural institutions and objects: "Each party to the conflict must protect cultural property: a) All seizure of or destruction or willful damage done to institutions dedicated to religion, charity, education, the arts and sciences, historic monuments and works of art and science is prohibited. b) Any form of theft, pillage or misappropriation of, and any acts of vandalism directed against, property of great importance to the cultural heritage of every people is prohibited" (ICRC 2018, Rule 40). Once again, and consistent with the IHL's anthropocentric frame of reference, these prohibitions are instituted for the sake of protecting the human population and their interests, not the object per se.

Instrumental Exclusions

The existing rules, regulations, and protocols of IHL are very clear. What matters, in all cases and contexts, are human beings; objects do not. Or, if they matter, they matter only to the extent that they belong to human beings, are necessary for the sake of human survival and well-being, or are recognized as important to the current and future state of human populations. This anthropocentric framework is entirely understandable insofar as it is consistent with the answer that is typically provided for the question concerning technology. "We ask the question concerning technology," Martin Heidegger (1977, 4–5) writes, "when we ask what it is. Everyone knows the two statements that answer our question. One says: Technology is a means to an end. The other says: Technology is a human activity. The two definitions of technology belong together. For to posit ends and procure and utilize the means to them is a human activity." According to Heidegger's analysis, the presumed role and function of any kind of technology—whether it be a simple hand tool, a jet airliner, or a sophisticated robot—is that it is a means employed by human users for specific ends. Heidegger (1977, 5) calls this particular characterization "the instrumental and anthropocentric definition" and indicates that it forms what is considered to be the "correct" understanding of any kind of technological contrivance.

As Andrew Feenberg (1991, 5) summarizes it, "The instrumentalist theory offers the most widely accepted view of technology. It is based on the common sense idea that technologies are 'tools' standing ready to serve the purposes of users." And because a tool or instrument is deemed "neutral," without intrinsic value or content of its own, a technological artifact is evaluated not in and of itself, but on the basis of the particular human and humane employments that have been decided by its designers, manufacturers, or users. Consequently, technology is only a means to an end; it is not and does not have an end in its own right. "Technical devices," as Jean-François Lyotard (1984, 33) writes, "originated as prosthetic aids for the human organs or as physiological systems whose function it is to receive data or condition the context. They follow a principle, and it is the principle of optimal performance: maximizing output (the information or modification obtained) and minimizing input (the energy expended in the process). Technology is therefore a game pertaining not to the true, the just, or the beautiful, etc., but to efficiency: a technical 'move' is 'good' when it does better and/or expends less energy than another." According to the instrumentalist way of thinking, a technological device, whether it be a corkscrew, a clock, or a LAWS, is a mere instrument of human action. The artifact does not in and of itself participate in the big and important questions of truth, justice, or beauty. It is simply and indisputably about efficiency. A particular technological innovation is considered "good" if and only if it proves to be a more effective instrumental means to accomplishing a humanly defined end.

Nonhuman Combatants

As Nowrot (2015, 135) has noted: "The scope of application of the *ius in bello*—as strongly indicated by its more recent labeling as international 'humanitarian' law—has always been and continues to be exclusively human-oriented." Although this way of thinking is entirely serviceable, it has increasingly become a problem for responding to and contending with nonhuman combatants.

Animal Soldiers

Excluded from IHL protections are animals that participate in combat operations, for example, dogs, horses, dolphins, sea lions, and so forth. But animals—and especially dogs—have been (and for quite some time now) something *more* than military equipment. First, animals already occupy a curious and ambivalent position with respect to human society both on and off the battlefield. "The way society regards animals," as Carpenter (2015, 43) writes, "can appear to be contradictory. The animals or pets in our home are considered family, but outside the home our relationships with animals change (e.g. work, feral, and animals raised for food). In other words, some animals are treated almost like people while others are used as tools."[2]

The consequences of this ambivalent status can be seen in the way canine assistants have been integrated into military contexts. According to US federal law, military working dogs (MWDs) are officially defined and classified as "military equipment." This decision, as Carpenter (2015, 43) points out, "comes from the necessary defaulting between only two choices the military currently assigns assets: humanpower or equipment." The classification of MWDs as equipment was legally codified by the passage of the Federal Property and Administrative Services Act of 1949. "One of the purposes of the Act," as Sarah Cruse (2015, 257) points out, "was to provide an economical and efficient system for the disposal of government surplus property." Despite this, soldiers in the field, and especially those individuals working closely with the dogs as handlers, "forge strong emotional bonds with these canines that are acting as part of their team regardless of the formal 'equipment' classification" (Cruse 2015, 257). As Nowrot (2015, 131) reports, "Joint participation in armed conflicts not infrequently results in the forming of rather close emotional bonds between human soldiers and their animals. They find their visible expression for example in the countless well-known stories about animals risking their lives to save human 'comrades' in critical situations; but there are also comparable tales about human soldiers doing the same for their canines, horses and other animals assigned to assist them on the battlefield." And the list of publications recounting the importance of these human–animal relationships and celebrating the valor of these animals in battle is impressive (Cooper 2000; Frankel 2016; Hediger 2013; Le Chêne 2010).

Second, many nations (including many North American and European countries) award both service medals and rank to animals. "The British," as Steven Johnston (2012, 368–69) recounts, "decorate animals. The Dickin Medal, named after the founder of the People's Dispensary for Sick Animals, Maria Dickin, functions as the animal version of the Victoria Cross. Animals receive plaudits for bravery, courage, and devotion to duty. The Medal reads 'For Gallantry, We also Serve,' suggesting patriotic oneness between human and animal. World War II saw 49 medals awarded." The US military not only bestows rank on its "animal soldiers," but in some circumstances have awarded MWDs military rank "that make them senior to their handlers, a practice designed to ensure that the humans treat the animals with deference" (Londono 2014). And several countries have now dedicated public memorials to honor animals for their combat service. As Nowrot (2015, 132) recounts: "On 21 July 1994, on the fiftieth anniversary of the invasion of Guam, the Marine War Dog Memorial was unveiled at the United States Marine Corps War Dog Cemetery on the island, dedicated to the twenty-five dogs killed 'liberating Guam in 1944.' To mention but one further example, the Animals in War Memorial was unveiled in London on 24 November 2004."

Finally, and because of this, there has been considerable effort to reclassify military animals. Cruse (2015), for instance, argues that the current classification of MWDs as equipment "grossly underestimates their role within the US military and deprives these dogs of the opportunity to transition to a peaceful civilian life once they are deemed 'excessive equipment' and retired from service." In response to this, as Carpenter (2015, 43) has reported, "there is a passionate MWD advocacy movement proposing to change the military classification for working canines from *equipment* to *manpower* (or a third, as yet undesignated category) in order to initiate and clarify the policies for prolonged care and maintenance of the dogs after retirement." And there has been some traction with efforts to alter the current law. "In February 2012 two identical bills titled the Canine Members of the Armed Forces Act were introduced in the [US] House and Senate to address the current status and treatment of MWDs. The purpose of the Canine Members of the Armed Forces Act was to reclassify MWDs as canine members of the armed forces, not equipment" (Cruse 2015, 264–265). Although both bills failed to garner the votes necessary to become law, their mere introduction demonstrates public recognition of the need to reconsider and reclassify MWDs.

Robot Soldiers

There are similar opportunities and challenges experienced in the face (or the face plate) of battlefield robots, which are also classified as military equipment. As Peter W. Singer (2009), Joel Garreau (2007), and Julie Carpenter (2015) have reported, soldiers have formed surprisingly close personal bonds with their units' explosive ordinance disposal (EOD) robots, giving them names, awarding them battlefield promotions, risking their own lives to protect that of the robot, and even mourning their death. This happens not because of how the robots are designed or what they are. It happens as a by-product of the way the mechanisms are situated within the unit and the role they play in battlefield operations. As Eleanor Sandry (2015, 340) explains:

> EOD robots, such as PackBots and Talons, are not humanlike or animal-like, are not currently autonomous and do not have distinctive complex behaviours supported by artificial intelligence capabilities. They might therefore be expected to raise few critical issues relating to human–robot interaction, since communication with these machines relies on the direct transmission of information through radio signals, which have no emotional content and are not open to interpretation. Indeed, the fact that these machines are broadly not autonomous precludes them from being discussed as social robots according to some definitions. . . . In spite of this, there is an increasing amount of evidence that EOD robots are thought of as team members, and are valued as brave and courageous in the line of duty. It seems that people working with EOD robots, even though the robots are machinelike and under the control of a human, anthropomorphise and/

> or zoomorphise them, interpreting them as having individual personalities and abilities.

There are three important items to note in this context. First, the robots that are profiled here look like and are deliberately designed to function as tools or military equipment. Existing EOD robots, like the PackBots and Talons mentioned by Sandry, are not designed for nor do they function as "social robots." Unlike the Furbie, the Pleo dinosaur, Pepper, or other sociable robots that are, as Cynthia Breazeal (2004) describes it, intentionally created for human social interaction, these battlefield robots are industrial looking and are created and deployed for the sole purpose of instrumental utility in the disposal of explosive ordinance. Unlike social robots that are fabricated to produce anthropomorphic projection (Darling 2016), the design, look, and function of EOD robots adhere to Joanna Bryson's (2010, 63) thesis that robots be built, marketed, and considered legally as serviceable tools (or what she calls, in a rather controversial choice of words, "slaves"), not companion peers.

Second, these EOD robots are not autonomous and in many cases would not even qualify as semiautonomous devices by any stretch of the imagination. Most of these devices are still under human remote control, do not incorporate or contain anything approaching advanced AI capabilities, and in most cases would not even be considered "smart technologies." Consequently existing EOD robots are considerably less capable than what is predicted for LAWS or fully autonomous killing machines. Despite this limitation, soldiers working with EOD robots respond to the mechanism *as if* they were another autonomous subject and not just a piece of equipment or mere object.

Finally, none of this is necessarily new or surprising. Evidence of this kind of response was already tested and demonstrated with Fritz Heider and Mariane Simmel's "An Experimental Study of Apparent Behavior" (1944), which found that human subjects tend to attribute motive and personality to simple animated geometric figures. Similar results have been obtained by way of the computer as social actor studies conducted by Byron Reeves and Clifford Nass in the mid-1990s. As Reeves and Nass discovered across numerous trials with human subjects, users (for better or worse) have a strong tendency to treat socially interactive technology, no matter how rudimentary, as if they were other people. "Computers, in the way that they communicate, instruct, and take turns interacting, are close enough to human that they encourage social responses. The encouragement necessary for such a reaction need not be much. As long as there are some behaviors that suggest a social presence, people will respond accordingly. When it comes to being social, people are built to make the conservative error: When in doubt, treat it as human. Consequently, any medium that is close enough will get human treatment, even though people know it's foolish and even though they likely will deny it afterwards" (Reeves and Nass 1996, 22).

EOD robots are designed to be tools. They are officially classified as military equipment. And all of this is entirely transparent and clearly communicated to users. Despite this, soldiers in the field often treat these robots as comrades-in-arms and not as a tool or just another piece of military hardware. They do so not because of what these robots *are* but because of the role that these artifacts play in combat operations and how they participate in or contribute to unit cohesion. This demonstrates that external, social-relational aspects, as both Mark Coeckelbergh (2012) and I (Gunkel 2017) have argued, often take precedence over and can be more determinative of actual social status and position than ontological properties or design intentions. Consequently, it is not necessarily what the robot is, how it has been designed, or to what extent its machinic nature is clearly communicated to users of the device that matters. What matters—and what seems to matter most on the battlefield, in particular—is how these robots comes to be situated within the unit, what functions they perform in combat operations with human partners, and what actually transpires between human soldiers and these artifacts in day-to-day interactions. In other words, what actually happens with and in the face of battlefield robots—as documented in the popular, military, and academic publications on the subject—often contravene and even undermine the best of design intentions, specifications, or rules concerning their official classification.

The Difference That Makes a Difference

Even though human–animal teamwork in battlefield operations can be looked at as a possible model for insight concerning human–robot relationships (Billings et al. 2012; Carpenter 2015, 42), these two nonhuman entities are not necessarily nor should they be considered *the same*.[3] On the one hand, animals have both a historic and biological privilege over robotic artifacts. Historically animals were here first. There were animals before there were robots, and animals have accompanied humans onto the battlefield from the earliest recorded accounts of human armed conflict. So there is, for lack of a better description, a kind of "first come first serve" logic that is operative in these discussions and debates. Biologically, human and nonhuman animals are the product of coevolution and have substantial similarities that go all the way down to the molecular level. It is not insignificant that *homo sapiens* has been estimated to share 97+ percent of its DNA with chimpanzees. Consequently, it seems entirely reasonable to address the opportunities and challenges made available by "animal soldiers" before attempting to take on and deal with the prospect of killer robots.

On the other hand, there are important ways that robots have a kind of precedence over animals, especially in situations regarding armed conflict. As Nowrot (2015, 140) points out, the status of "combatant," as it is currently described under IHL, "involves not only the enjoyment of certain rights and privileges but also—at least indirectly—an imposition of certain legal obligations." And because it is doubtful whether an animal, like a MWD, would be capable of performing these obligations, it seems unlikely that a reclassification of animals as a kind of "combatant" would be successful. As Nowrot (2015, 141) concludes: "A sober evaluation—to put it mildly—gives rise to certain doubts of whether or not the average animal combatant can seriously be regarded as being endowed with the capacity to understand and autonomously obey the various legal obligations incumbent upon active participants in armed conflicts under international humanitarian law." For instance, it not just unlikely but virtually impossible to expect that an animal would be capable of the obligation imposed by IHL on combatants to have "the capacity to distinguish persons who participate in the hostilities from those who do not or to make proportionality calculations" (de Hemptinne 2017, 274).

Unlike animals, however, robots can be (or at least have a higher likelihood of being) programmed to follow the rules of IHL—assuming that the rules, like that involving the calculation of proportionality, could be made computable (which, for now, remains an open question). In fact, as Ronald Arkin (2009) argues, it may be the case that robots will be better at following the rules of military engagement than fallible human soldiers. Arkin, in fact, lists six reasons why autonomous robots "may be able to perform better than humans" in the "fog of war," including the following: (1) Robots do not need "to have self-preservation as a foremost drive" and therefore "can be used in a self-sacrificing manner if needed." (2) Machines can be equipped with better sensors that exceed the limited capabilities of the human faculties. (3) "They can be designed without emotions that cloud their judgment or result in anger and frustration with ongoing battlefield events." And (4) "they can integrate more information from more sources far faster before responding with lethal force than a human possibly could in real-time" (Arkin, 2009, 29–30).

Additionally, and precisely because of this, advocates for efforts to reclassify "animals soldiers" in both national laws and international agreements now recognize that working on the "machine question" might have a discursive advantage and open a "window of opportunity" for addressing the status of the animal. As Nowrot (2015, 144) candidly points out at the end of his essay, "International humanitarian law and its lawmaking actors are in the foreseeable future highly likely to be confronted with the question of how to legally cope with a number of other categories of 'nonhuman combatants,' particularly in the form of autonomous combat systems." And "it is precisely under such circumstances, one could argue, that the international community might also be more willing to discuss a possible legal status for animal soldiers within a more comprehensive reformation of the laws of war in general and the scope of this normative regime's application in particular" (Nowrot 2015, 145). In other words, the animal question might only get traction in the wake of developments with LAWS such that contending with the opportunities and challenge of the "machine question" paves the way for articulating and responding to the reclassification of animals. In terms of rhetorical strategy, then, the best way to respond to the needs of animals might be to work on behalf of the robot.

Solutions and Outcomes

When it comes to nonhuman combatants, the existing documents and rules governing international armed conflict appear to be out of sync with actual practices on the ground. Despite the fact that both animals and robots are officially classified as military equipment that can be used and even abused with very little moral or legal restrictions, these nonhuman "others" occupy positions that, for better or worse, make them something more than a mere object. They may not ever be as "valuable" as other persons—whether human combatants or civilians—but there is considerable and documented resistance to reducing them to the status of a mere tool or piece of military hardware. So how can or should we respond to this opportunity/challenge? Let me conclude by considering two different and opposed outcomes.

Status Quo

We can continue to deploy and enforce the existing rules of the game. International humanitarian law is, in both name and substance, "deeply anthropocentric" (de Hemptinne 2017, 272). It therefore, as we have seen, stipulates the lawful conduct of human activity during armed conflict for the sake of respecting the dignity and well-being of human persons. All other nonhuman entities involved in these actions—animals, the natural environment, cultural objects, and machines—are and remain nothing more than things and equipment serving the interests of human persons and communities. This way of thinking—this way of dividing between *persons* who count and *things* that do not—clearly works. It is, one might say, of considerable instrumental utility when confronting and deciding matters of rights and responsibilities in the unfortunate situations of international armed conflict.

A good example of this kind of reassertion of the status quo can be found in recent international efforts to ban killer robots. In 2009, Jürgen Altmann, Peter Asaro, Noel Sharkey, and Robert Sparrow organized the International Committee for Robot Arms Control (ICRAC), calling "upon the international community for a legally binding treaty to prohibit the development, testing, production and use of autonomous weapon systems in all circumstances" (ICRAC 2017). In 2013, ICRAC partnered with sixty-three other international and national NGOs from twenty-eight countries in the Campaign to Stop Killer Robots. In effect, the campaign has argued that advanced weapon systems, no matter how sophisticated their design or operations, must always be tethered to and remain under human control, and there should always be a human being in the loop who is able to take responsibility and to be held accountable for targeting and attack decisions. The Berlin Statement from ICRAC (2017) advances a similar cause: "We believe that it is unacceptable for machines to control, determine, or decide upon the application of force or violence in conflict or war. In all cases where such a decision must be made, at least one human being must be held personally responsible and legally accountable for the decision and its foreseeable consequences."

But this top-down reassertion of technological instrumentalism, whereby battlefield robots are declared to be instruments of human decision-making and action, is increasingly challenged and put in question by moral intuitions and military practices that appear to proceed otherwise—that consider the operational status of nonhuman combatants in ways that are both different from and potentially abrasive to this essentially instrumentalist way of thinking. Even if the campaign is successful in restricting battlefield robots to the category of equipment—ostensibly a tool that is always under the direct control and oversight of a human operator—the role that these machines already play in actual combat operations effectively challenge this formulation. Now please understand what is being argued here. I am not taking up a position that is simply opposed to the campaign's mission or advocating for the design and indiscriminate deployment of LAWS. What is being argued is that efforts to restrict LAWS—like that introduced and promoted by the Campaign to Stop Killer Robots—may need to be reformulated in a way that is more sensitive and attentive to the actual situations regarding battlefield robots (and animals[4]), recognizing that strict imposition of anthropocentric instrumentalism, although entirely workable in theory, risks being "tone deaf" to actual, documented practices.

Reclassification

Instead of reasserting a strict form of anthropocentric instrumentalism, there may be something to gain from recognizing that actual practices in the field do not (and perhaps never really did) adhere to the simple ontological dichotomy that divides the world of entities into persons or equipment. Evidence from "boots on the ground" experience demonstrate that both animal soldiers and robot soldiers, like MWDs and EODs, have occupied and continue to occupy liminal positions that do not fit this neat and somewhat artificial categorization. And in response to this, there may be good reasons to recategorize these nonhuman combatants as something other than military equipment.

But one needs to be attentive to what is actually being proposed here. No one is suggesting that military animals or killer robots be classified as personnel on par with human beings—combatant, person hors de combat, or civilian. What is argued is that the existing classification schema—one that recognizes only two kinds of entities, personnel or equipment—may simply be too restrictive and insensitive to respond to and take responsibility for the different kinds of things with which we interact and involve ourselves. "In light of these differences," Nowrot (2015, 143) argues, "it appears appropriate and advisable from a legal policy perspective not to transfer and extend the current concept of (human) combatants 'lock, stock, and barrel' to animal soldiers but rather create a new separate category of animal combatants under international humanitarian law."[5] Something similar can and perhaps should be explored for the sake of "robot soldiers." Although currently existing battlefield robots, like the very industrial-looking and utilitarian EODs, are not sentient or even significantly autonomous (by any definition), do not have specific interests (as far as we know), and cannot be harmed in any appreciable way (at least not beyond what is typically regarded as property damage), this does not mean that they do not or should not have some level of recognition for the sake of responding to and respecting the human–robot teams in which they are situated and from which they operate.

This reclassification of things, especially as it applies to battlefield robots, would not, it is important to point out, simply invalidate the instrumentalist theory *tout court*. There are and will continue to be things that are and function as military equipment. But it does mean that everything is not and should not be treated the same. In addition to the usual stock of military equipment, such as rifles, tanks, tents, radio transmitters, and ammunition, there are also other kinds of things—animals and robots. Unlike mere objects, these animal and robot "soldiers" have a very real social presence that matters for the individual human beings who work with them and the social formations in which they have been situated and function. Things, therefore, do not participate in a "flat ontology." There are differences. And our moral and legal systems—even and especially in the case of armed conflict—should be more fine-grained and able to resolve and contend with these differences. The division of battlefield assets into the two exclusive categories of personnel or equipment is imprecise, inflexible, and potentially inhumane.

Notes

1. The items listed here—privileges, claims, powers, and immunities—comprise the four Hohfeldian incidents that define and characterize moral, legal, and political rights. For more on the Hohfeldian typology, see Wenar (2005).

For a consideration of the way the Hohfeldian incidents apply to and explain rights in the context of AI and robotics, see Gunkel (2018).

2. Mark Coeckelbergh and I have examined this curious and seemingly contradictory situation regarding the social status of nonhuman animals in "Facing Animals: A Relational, Other-Oriented Approach to Moral Standing" (Coeckelbergh and Gunkel 2014).

3. This is an important point made by Katharyn Hogan (2017) in an essay that was published as a critical rejoinder to *The Machine Question* (Gunkel 2012). In the essay, Hogan argues, in opposition to what she presumes had been developed in the book, that "the machine question is not the same question as the animal question" (Hogan 2017, 29). Hogan is absolutely correct about this. The question concerning the machine and the question concerning the animal are, in fact, not the same question. But she is incorrect in assigning this (mistaken) position to the book, which developed a much more nuanced treatment of things that is attentive to difference. Consequently, Hogan's essay is at best half right and can only proceed and succeed by way of formulating what amounts to a straw man argument.

4. This parenthetical addition concerning animals is not insignificant. The Campaign to Stop Killer Robots, as its name indicates, is exclusively concerned with technological artifacts; it has had little or nothing to say about animals. But LAWS could also be developed on an animal rather than a technological platform. "Using more advanced animals as cyborgs," as Armin Krishnan (2017, 74) has argued, "is clearly possible. Apes could be engineered as animal soldiers: they could be made smarter by adding some human DNA that enhances their cognitive abilities to a certain extent. If more than 50 per cent was animal DNA, they would probably not be granted human rights status and could be used as slaves or as expendable fighters."

5. This has in fact transpired in some militaries, where animal soldiers are deliberately not classified as equipment in order to enable their human handlers to adopt the animals after they are retired from military service. Whether this status would extend to and be recognized outside this particular context, that is, if the animal were captured by the enemy, is another matter.

References

Arkin, Ronald. 2009. *Governing Lethal Behavior in Autonomous Robots*. Boca Raton, FL: Chapman and Hall/CRC.

Asaro, P. 2012. "On Banning Autonomous Weapon Systems: Human Rights, Automation, and the Dehumanization of Lethal Decision-Making." *International Review of the Red Cross* 94(886): 687–709. https://international-review.icrc.org/articles/banning-autonomous-weapon-systems-human-rights-automation-and-dehumanization-lethal.

Billings, Deborah R., Kristin E. Schaefer, Jessie Y. C. Chen, Vivien Kocsis, Maria Barrera, Jacquelyn Cook, Michelle Ferrer, and Peter A. Hancock. 2012. "Human-Animal Trust as an Analog for Human-Robot Trust: A Review of Current Evidence." Technical Report ARL-TR-5949. Orlando, FL: Army Research Laboratory. https://www.arl.army.mil/arlreports/2012/ARL-TR-5949.pdf.

Breazeal, Cynthia L. 2004. *Designing Sociable Robots*. Cambridge, MA: MIT Press.

Bryson, Joanna. 2010. "Robots Should Be Slaves." In *Close Engagements with Artificial Companions: Key Social, Psychological, Ethical and Design Issues*, edited by Yorick Wilks, 63–74. Amsterdam: John Benjamins.

Campaign to Stop Killer Robots. 2017. Main Page. http://www.stopkillerrobots.org.

Carpenter, Julie. 2015. *Culture and Human-Robot Interaction in Militarized Spaces: A War Story*. New York: Ashgate.

Coeckelbergh, Mark. 2012. *Growing Moral Relations: Critique of Moral Status Ascription*. New York: Palgrave Macmillan.

Coeckelbergh, Mark, and David J. Gunkel. 2014. "Facing Animals: A Relational, Other Oriented Approach to Moral Standing." *Journal of Agricultural & Environmental Ethics* 27(5): 715–33. https://doi.org/10.1007/s10806-013-9486-3.

Cooper, Jilly. 2000. *Animals in War*. London: Corgi Books.

Cruse, Sarah D. 2015. "Military Working Dogs: Classification and Treatment in the U.S. Armed Forces." *Animal Law* 21(2): 249–84. https://law.lclark.edu/live/files/23698-21-crusepdf.

Darling, Kate. 2016. "Extending Legal Protection to Social Robots: The Effects of Anthropomorphism, Empathy, and Violent Behavior toward Robotic Objects." In *Robot Law*, edited by Ryan Calo, A. Michael Froomkin, and Ian Kerr, 213–31. Northampton, MA: Edward Elgar.

de Hemptinne, Jérôme. 2017. "The Protection of Animals during Warfare." *American Journal of International Law* 111(1): 272–76. https://doi.org/10.1017/aju.2017.69.

Derrida, Jacques. 2008. *The Animal That Therefore I Am*. Translated by David Wills. New York: Fordham University Press.

Feenberg, Andrew. 1991. *Critical Theory of Technology*. Oxford: Oxford University Press.

Frankel, Rebecca. 2016. *War Dogs—Tales of Canine Heroism, History, and Love*. New York: St. Martin's Press.

Garreau, Joel. 2007. "Bots on the Ground: In the Field of Battle (or Even above It), Robots Are a Soldier's Best Friend." *Washington Post*, May 6. http://www.washingtonpost.com/wp-dyn/content/article/2007/05/05/AR2007050501009.html.

Gunkel, David J. 2012. *The Machine Question: Critical Perspectives on AI, Robots, and Ethics*. Cambridge, MA: MIT Press.

Gunkel, David J. 2017. "The Other Question: Can and Should Robots Have Rights?" *Ethics and Information Technology*. https://doi.org/10.1007/s10676-017-9442-4.

Gunkel, David J. 2018. *Robot Rights*. Cambridge, MA: MIT Press.

Hediger, Ryan. 2013. *Animals and War: Studies of Europe and North America*. Boston: Brill.

Heidegger, Martin. 1977. *The Question concerning Technology and Other Essays*. Translated by W. Lovitt. New York: Harper & Row.

Heider, Fritz, and Marianne Simmel. 1944. "An Experimental Study of Apparent Behavior." *American Journal of Psychology* 57(2): 243–59. https://doi.org/10.2307/1416950.

Hogan, Katharyn. 2017. "Is the Machine Question the Same Question as the Animal Question?" *Ethics and Information Technology* 19(1): 29–38. https://doi.org/10.1007/s10676-017-9418-4.

ICRC. 2018. "International Committee of the Red Cross." IHL Database/Customary IHL. https://ihl-databases.icrc.org/customary-ihl/eng/docs/home.

International Committee for Robot Arms Control (ICRAC). 2017. About ICRAC. https://icrac.net/.

Johnson, Aaron M., and Sidney Axinn. 2013. "The Morality of Autonomous Robots." *Journal of Military Ethics* 12(2): 129–41. https://doi.org/10.1080/15027570.2013.818399.

Johnston, Steven. 2012. "Animals in War: Commemoration, Patriotism, Death." *Political Research Quarterly* 65(2): 359–71. http://www.jstor.org/stable/41635239.

Krishnan, Armin. 2013. *Killer Robots: Legality and Ethicality of Autonomous Weapons*. New York: Routledge.

Krishnan, Armin. 2017. *Military Neuroscience and the Coming Age of Neurowarfare*. New York: Routledge.

Le Chêne, Evelyn. 2010. *Silent Heroes: The Bravery and Devotion of Animals in War*. London: Souvenir Press.

Levy, David. 2005. *Robots Unlimited: Life in a Virtual Age*. Boca Raton, FL: CRC Press.

Lokhorst, Gert-Jan, and Jeroen van den Hoven. 2012. "Responsibility for Military Robots." In *Robot Ethics: The Ethical and Social Implications of Robots*, edited by K. Abney, P. Lin, and G. A. Bekey, 145–56. Cambridge, MA: MIT Press.

Londono, Ernesto. 2014. "Military Dog Captured by Taliban Fighters, Who Post Video of Their Captive." *Washington Post*, February 6. https://www.washingtonpost.com/world/national-security/military-dog-captured-by-taliban-fighters-who-post-video-of-their-captive/2014/02/06/c8d0f8f0-8f44-11e3-84e1-27626c5ef5fb_story.html?utm_term=.7429801a39d2.

Lyotard, Jean-François. 1984. *The Postmodern Condition: A Report on Knowledge*. Translated by G. Bennington and B. Massumi. Minneapolis: University of Minnesota Press.

Noorman, Merel, and Deborah G. Johnson. 2014. "Negotiating Autonomy and Responsibility in Military Robots." *Ethics and Information Technology* 16(1): 51–62. https://doi.org/10.1007/s10676-013-9335-0.

Nowrot, Karsten. 2015. "Animals at War: The Status of 'Animal Soldiers' under International Humanitarian Law." *Historical Social Research* 40(4): 128–50. https://doi.org/10.12759/hsr.40.2015.4.128-150.

Reeves, Byron, and Clifford Nass. 1996. *The Media Equation: How People Treat Computers, Television, and New Media Like Real People and Places*. Cambridge: Cambridge University Press.

Sandry, Eleanor. 2015. "Re-Evaluating the Form and Communication of Social Robots: The Benefits of Collaborating with Machinelike Robots." *International Journal of Social Robotics* 7(3): 335–46. https://doi.org/10.1007/s12369-014-0278-3.

Sharkey, Noel. 2012. "Killing Made Easy: From Joysticks to Politics." In *Robot Ethics: The Ethical and Social Implications of Robots*, edited by K. Abney, P. Lin, and G. A. Bekey, 111–28. Cambridge, MA: MIT Press.

Singer, Peter Warren. 2009. *Wired for War: The Robotics Revolution and Conflict in the Twenty-First Century*. New York: Penguin Books.

Sparrow, Robert. 2007. "Killer Robots." *Journal of Applied Philosophy* 24(1): 62–77. https://doi.org/10.1111/j.1468-5930.2007.00346.x.

Wenar, Leif. 2005. "The Nature of Rights." *Philosophy & Public Affairs* 33(3): 223–52. http://www.jstor.org/stable/3557929.

Chapter 6

No Hands or Many Hands?

Deproblematizing the Case for Lethal Autonomous Weapons Systems

JAI GALLIOTT

Central to the ethical concerns raised about the development of increasingly intelligent autonomous weapons systems are issues of responsibility and accountability.[1] Robot arms control groups have popularized this element of the debate as part of their call for a moratorium on the use of autonomous military robotics. The purpose of this chapter is to demonstrate that, while autonomous systems certainly exacerbate some traditional problems and may in some cases cause us to rethink who we ought to hold morally responsible for military war crimes, our standard conceptions of responsibility are capable of dealing with the supposed "responsibility gap"—namely the inability to identify an appropriate locus of responsibility—in autonomous warfare and that, in the absence of a gap, there is no reason for an outright ban on responsibility attribution or human control grounds (see Galliott 2013a, 2013b, 2016, 2017).

This chapter begins by exploring the basis for the attribution of responsibility in just war theory. It then looks at the conditions under which responsibility is typically attributed to humans and how these responsibility requirements are challenged in technologically mediated warfare. Following this is an examination of Robert Sparrow's notion of the "responsibility gap" as it pertains to the potential deployment of fully autonomous weapons systems. It is argued that we can reach a solution by shifting to a forward-looking and functional sense of responsibility, which incorporates institutional agents and ensures that the human role in both engineering and releasing these systems is never overlooked.

Responsibility in War

In *Just and Unjust Wars*, Walzer (1977) writes that the "assignment of responsibility is . . . critical" because "there can be no justice in war if there are not, ultimately, responsible men and women." This is a logical claim as any defensible moral theory ought to incorporate a theory of moral responsibility or call for people to explain their intentions and beliefs if it is to lead to the rectification of wrongs. However, many theorists have simply asserted the importance of the assignment of responsibility in just war theory and proceeded to argue that the challenges presented for this assignment to intelligent systems (which can autonomously apply lethal force) are insurmountable, and so such systems should therefore be banned (International Committee for Robot Arms Control 2013; Keller 2013). Sparrow (2007), for instance, argues that it is a fundamental condition of fighting a just war that "someone" be held responsible for any civilian deaths or other unjust actions (Sparrow 2007). He goes on to say that the assumption or allocation of responsibility is also "vital in order for the principles of *jus in bello* to take hold at all" (Sparrow 2007). Royakkers and van Est (2010) take Sparrow's argument to mean that it is a condition of just war theory that an "individual" be able to be held responsible. One strong implication of the justice of warfare being separate from the justice of the war itself is that it permits the judging of acts within war to be disassociated from war's cause, even if all ties between cause and justice in war are not completely severed (Fieser and Dowden 2007). This is a useful division when it comes to examining Sparrow's former claim about the merit of *jus in bello* principles, though we first need to clarify the extent to which responsibility and accountability are necessary conceptual conditions for systematic compliance with the moral laws and customs of war.

When we speak of the necessity of attributing responsibility in this context, we are speaking of the need for accountability for harm or death caused in the targeting process and the subsequent decision to release (or not to release) deadly munitions. We know from just war theory that it is considered impermissible to attack people indiscriminately since noncombatants are considered outside the bounds of the battlefield. Their immunity in war stems from the fact that their existence and activity has little to do with war, with combatant status typically reserved for those involved in the prevention of harm and the rectification of wrongs through killing as agents acting on behalf of their respective states. Whoever or whatever is tasked with making targeting decisions must therefore proffer a reason as to why combatants become appropriate targets in the first instance. The difficulty of this task varies depending on the standards to which war-making agents are held. Theorists have traditionally invoked rights analyses in order to give soldiers an idea of who is an appropriate target, though we will not go into these here. To build again on what was said earlier, some also hold that simply being trained and armed constitutes a sufficient justification and that the donning of a uniform and the carrying of arms signifies a change in moral status of the person and a shift in their status from noncombatant to combatant. Others invoke the boxing ring analogy. This analogy holds that, while punching a man on a street in a civilized society is unfair and immoral, fighting those who voluntarily step into the ring is perfectly justifiable. These "voluntarists" imply that soldiers renounce their immunity and become legitimate targets once they set foot on the battlefield (McMahan 2004). Irrespective of the underlying rationale, the task of discrimination is made much more onerous because of the character of modern warfare and of the role that civilians play in aiding technological warfare and the wider "war machine" (McMahan 2004).

However, the point here is that no matter how difficult the distinction between combatant and noncombatant is to make in practice, a lack of identifiability does not give a government or its agents the right to kill indiscriminately. The onus is very much placed on the active aggressor to identify combatants and take "due care" in the process. The implication, then, is that if there is a particular degree of uncertainty, an attack should not go ahead. If it does go ahead and civilians are killed, responsibility for the attack must generally rest upon the shoulders of those who made the decision. This assumes that we can identify participants in the targeting decision. The same holds true when it comes to tempering the extent and violence of warfare and minimizing general destruction. While the discrimination requirement discussed above is concerned with who or what is a legitimate target of war, the proportionality requirement deals more explicitly with how much and what sort of force is morally permissible to use in seeking to destroy those targets. Even in a battle that involves military targets only (a rare occurrence in modern times), it is possible to employ proportionate force, since one is morally sanctioned to use only the amount of force necessary to achieve the objective at hand (Walzer 2006). Of course, many believe that autonomous systems allow their users to avoid some of war's suffering by employing highly selective killing. If a small number of legitimate targets can be killed as part of a military operation to avoid further bloodshed, then this makes a great deal of sense according to proportionality. If, however, it brings civilians and their infrastructure into the line of danger, then it may not be so wise to employ such practices. It will only complicate targeting processes and responsibility ascription.

In any case, the principles of discrimination and proportionality play an important role in aiming to restrict war's violence and range, but there is no obvious reason why an *individual person* ought to take the blame where collective violations of these principles have occurred, as per the Royakkers and van Est interpretation. Surely they mean to say that these principles underwrite the requirement that *agents of war*, whether human, machine, or some combination thereof, should be held responsible for their actions. Of the many reasons for this, three stand out. The first, and probably the most important, is that states and their agents need to learn from their mistakes and those of others in order to ensure that future wars achieve better results and are conducted more justly (Walzer 2006). The second claims that the majority of people (especially the victims, their families, and the wider affected civilian or military community, or all three) will not find it satisfactory if, when a casualty occurs, there is nobody or nothing held responsible, that is, nothing toward which the victim's family and so forth can direct attitudes such as blame and resentment or compensation requests. Third, for soldiers, organizations, and the like, there is also a respect in which taking responsibility for the role they played in certain actions can help them transition back into the civilian realm with a good conscience or ethical reputation. After all, not all modern soldiers are permanent enlisted members of their respective defense forces that serve overseas, nor are all organizations permanent parts of the military–industrial complex. Therefore, even in peacetime, the right conception of responsibility acts as an inhibitor of immoral conduct in reasserting the importance of adhering to the laws/customs of war and reminding agents of war and those supporting them that war will eventually end and that they will want to return to civilian life without being plagued by problems that can arise from improper actions and conduct.

That said, when such actions occur, allocating responsibility is just as difficult as differentiating between combatant and noncombatant or making proportionality judgments. As Walzer himself admits, "it is no longer easy to impose responsibility," especially with "certain sorts of state action, secretly prepared or suddenly launched" (Walzer 2006). Thus, it should come as no surprise that many significant war crimes have gone without the assignment of responsibility that is necessary to accord with Walzer's just war theory. In some cases, allocating responsibility is so difficult that Walzer tries to avoid holding specific actors responsible despite his claims about the necessity of doing so. For example, he talks about the many Americans who "were morally complicitous in our Vietnam aggression," though he says shortly thereafter that he is not "interested in pointing at particular people or certain that he can do so" (Walzer 2006). Instead, he only wants to insist that "there are responsible people . . . [and that] the moral accounting is difficult and imprecise" (Walzer 2006). As will be demonstrated in the next section, this moral accounting is only made more difficult in the technological age. Royakkers and van Est suggest that this might not be such a big problem, though. They say that "we do not normally say that whether or not someone may be held responsible after the fact . . . is a condition of permissibility, at least in the absence of a rule to the contrary" (Royakkers and van Est 2010). In other words, they suggest that the permissibility of war is not tied to post-facto responsibility identification unless there is some rule establishing the link. However, war is unlike their purportedly normal circumstances. The requirements for waging war are notably different to those for resorting to violence in peacetime. They ignore the fact that there is a rule of responsibility implicit in just war theory and that, even if there were no rule, this would be a contributor to things like seemingly endless conflicts and thus an important reason to develop such a rule and investigate the challenges in adhering to it.

Challenges to Responsibility Attribution in Technologically Enabled Warfare

Moral responsibility in war is about actions, omissions, and their consequences. When we read stories in military ethics readers, those worthy of blame include agents failing to adhere to just war principles or to otherwise do the "right thing" by platoon leaders, government, or country (Walzer

2006). It is also about the conditions under which they did the right or wrong thing. To be held responsible in accord with Fischer and Ravizza's (1998) landmark account—which is based on the idea of guidance control and that the mechanism that issues the relevant behavior must be the agent's own and be responsive to reasons—actors must not be "deceived or ignorant" about what they are doing and ought to have control over their behavior in a "suitable sense" (Fischer and Ravizza 1998). Put more specifically, this means that an agent should only be considered morally responsible if they intentionally make a free and informed causal contribution to the act in question, meaning that they must be aware of the relevant facts and consequences of their actions, have arrived at the decision to act independently, and were able to take alternative actions based on their knowledge of the facts (Fischer and Ravizza 1998). If these conditions are met, we can usually establish a link between the responsible subject and person or object affected, either retrospectively or prospectively (the latter will be the focus of the final section). However, technologically enabled warfare of the autonomous type presents various challenges for these standard accounts of moral responsibility. For the sake of a complete exposition and refutation of Sparrow's claim that the responsibility gap presents an insurmountable threat, it is necessary to take a closer look at how semiautonomous military technologies, generally defined, can complicate responsibility attribution in warfare.

There are many barriers to responsibility attribution in the military domain and many are so closely interrelated that it makes providing a clear and lucid discussion quite problematic. The most important for present purposes is associated with the subject's causal contribution to the action in question. According to the above referred-to account, for an agent to be held responsible, they must have exerted due influence on the resulting event. What is "due" will be subject to further reflection in the remaining sections, but there is little to be gained from blaming someone or something for an unfortunate event about which they/it legitimately had no other choice or over which they/it had no control. That acknowledged, the employment of modern warfighting technologies based on complex computing and information technologies can lead us to lose our grasp of who is responsible, because it obscures the causal connections between an agent's actions in contributing to semiautonomous or autonomous warfare and the eventual consequences. When utilizing complex technologies, tracing the sequence of events that led to a particular event usually leads in a great number of directions (Waelbers

2009; Jonas 1984; Fischer and Ravizza 1993; Noorman 2012). The great majority of technological mishaps are the product of multifaceted mistakes commonly involving a wide range of persons, not limited to end users, engineers, and technicians. For those looking from the outside in, it can be very difficult—and some (like Sparrow) might say impossible—to identify contributing agents. This difficulty in identifying contributing agents is Dennis Thompson's (1980, 1987) so-called problem of many hands. This problem should not be confused with the "responsibility gap" which will soon be addressed, because it is not as deflationary and falls short of the complete abdication of responsibility.

Added to the problem of many hands is the physical distance that warfighting technologies often create between agents and the consequences or outcomes of their actions. This further blurs the causal connection between action and event. Batya Friedman (1990) earlier noted this effect in an educational setting that encourages young people to become responsible members of the electronic information community. The challenge has been reinvigorated with the development and deployment of autonomous systems in the military setting and the employment of distanced drone operators. It is these war-making agents that now need to be further encouraged to play a responsible role in network-centric operating environments and have their role fully acknowledged by states and those seeking a ban on autonomous technologies. Autonomous systems technologies—more than any other material technology—extend the reach of military activity through both time and space. While enabling a state's military force to defend itself over a greater range than they would otherwise be able to may be morally permissible, or even praiseworthy, and may be in line with the social contract, this remoteness can also act to disassociate them from the harm that they cause. It has long been understood that there is a positive relationship between the physical and emotional distance facilitated by technological artifacts and the subsequent ease of killing (Grossman 1995). When someone uses an autonomous aircraft operated from a control station on the ground in the United States to conduct military operations in the Middle East, the operator might not be fully aware of how the system and its munitions will affect the local people and may not experience or fully appreciate the true consequences of their actions, psychological or collateral (Waelbers 2009). This has a direct bearing on their comprehension of the significance of their actions and has a mediating role when it comes to considering the extent to which they are responsible.

This mediation of responsibility has much to do with the fact that autonomous systems and the sensors that they carry can actively shape how moral agents perceive and experience the world at large, which further affects the conditions for imposing moral responsibility. In order to make the appropriate decisions that are sanctioned by just war theory, a moral agent must be capable of fully considering and deliberating about the consequences of their actions, understanding the relevant risks and benefits they will have, and to whom they will apply. This, in turn, calls for them to have adequate knowledge of the relevant facts. While McMahan (2009) and others have offered accounts (Zuboff 1985), it remains unclear what epistemic thresholds ought to apply here, but what is generally accepted is that it is unfair to hold someone responsible for something they could not have known about or have reasonably anticipated. The capability of autonomous systems and other intelligence-gathering technologies is importantly relevant, because in some respects they assist the relevant users in deliberating on the appropriate course of action by helping them capture, collate, and analyze information and data (United States Department of Defense 2009). In their sales demonstrations to the military, for example, representatives of the drone industry typically argue that their piece of military hardware will grant them the opportunity to see "beyond the next hill" in the field and "around the next block" in congested urban environments, enabling them to acquire information that they would not otherwise have access to without incurring significantly greater risk (Cummings 2004). This may well be true with respect to some systems, and these would allow operators greater reflection on the consequences of their tactical decisions. However, with the technical, geographical, and operational limits imposed, there are many respects in which these systems preclude one from gaining a view of the "bigger picture" and may alter an operator's resulting actions, perhaps limiting responsibility.

Many intelligent military systems have such complex processes that they get in the way of assessing the validity and relevance of the information they produce or help assess; they can actually prevent a user from making the appropriate decision within an operational context and therefore have a direct impact on their level of responsibility. A consequence of this complexity is that people have the tendency to rely either too much or not enough on automated systems like those we increasingly find embedded in autonomous aircraft or their control systems, especially in the time-critical and dynamic situations that are characteristic of modern warfare. The USS *Vincennes* most shockingly illustrated this during its deployment to the Persian Gulf amid a gun battle with Iranian small boats. Although this warship was armed with an Aegis Combat System, which is arguably one of the most complex and automated naval weapons system of its time (it can automatically track and target incoming projectiles and enemy aircraft), the USS *Vincennes* misidentified an Iranian airliner as an F-14 fighter jet and fired upon it, killing nearly three hundred people (Campbell and Eaton 2011; Gray 1997). Postaccident reporting and analysis discovered that overconfidence in the abilities of the system, coupled with a poor human–machine interface, prevented those aboard the ship from intervening to avoid the tragedy. Despite the fact that disconfirming evidence was available from nearby vessels as to the nature of the aircraft, it was still mischaracterized as a hostile fighter descending and approaching them at great speed. In the resulting investigation, a junior officer remarked that "we called her Robocruiser. . . . she always seemed to have a picture and . . . [always] seemed to be telling people to get on or off the link as though her picture was better" (Rogers and Rogers 1992). The officer's impression was that the semiautonomous system provided reliable information that was otherwise unobtainable. In this case, at least, such a belief was incorrect. The system had not provided otherwise unobtainable information, but rather misleading information. It is therefore questionable whether the war-making agent has a more comprehensive understanding of the relevant state of affairs because of the employment of advanced military technology or whether his understanding and knowledge is less accurate (Manders-Huits 2006). That is, it is unclear whether the attribution of moral responsibility is enhanced or threatened. This challenge is not unique to autonomous systems or indeed technology. When somebody relies on another's information to provide context or information (say another soldier), there also exists a misinformation risk. Nevertheless, the view advanced here is that, even though there may be an aggregate increase in the amount of information that is accessible, there is a morally relevant decrease in understanding which single piece of information ought to influence autonomy of action and the resulting decision-making, even when the bulk of information is clear and accurate. The implication is that operators of sophisticated systems might be held to high standards of responsibility on the basis that they had access to a great deal of relevant information when, in fact, the provision of this information may have clouded their judgment, meaning that they are less responsible.

It must also be added that advanced technologies may exert a certain level of influence over their users in a way that might be unclear or even immeasurable (Manders-Huits 2006). Again, one might argue that there is little difference between this and the loss of control inherent in situations in which one human officer orders other human soldiers to perform a particular task or mission. Once those individuals depart on their own, the officer likewise does not have total control over how that task or mission is accomplished. However, the cases differ in that with technology, the system is less fluid, less a matter of natural social construct and more by artificial construct. While this sort of control is not implicit in the technology itself, it is rather exerted through the design process and the way in which alternative moral options are presented for human action. Semiautonomous military technologies help to centralize and increase control over multiple operations, reducing costs and supposedly increasing efficiency. However, there is a limit to how much control a human being can exert. In reality, this "increased control" can only be achieved by outsourcing some low-level decisions to computerized processes and leaving the human to make a choice from a more limited range of actions. In other words, some military technologies are designed with the explicit aim of making humans behave in certain ways, further mediating the imposition of responsibility. However, note that we are still a long way from saying that we cannot attribute responsibility in such cases.

The Alleged Responsibility Gap in Autonomous Warfare

In the previous section, we saw how developments in military technology have led to a partial loss of influence on the part of operators or users and hold broader implications for the attribution of moral responsibility more generally, namely by limiting the operator's responsibility and perhaps causing us to consider the redistribution of the remaining share of responsibility. In the following section, we shall see how many of the problems described above are only being exacerbated as autonomous systems become more computerized. These problems and others come together in fully automated warfare to create a problem that Sparrow and others see as significantly more serious than any of those already discussed, posing what is supposedly an insurmountable threat to the responsibility framework embedded within just war theory. As a final step toward fully understanding and refuting the nature and implications of this problem, it will be necessary to discuss the arguments drawn on by Sparrow, namely those originally put forward by Andreas Matthias (2004).

Matthias (2004) argues that the further we progress along the autonomy continuum, the closer we are to undermining society's centuries-old effort to establish rule systems in order to attribute responsibility. He says that with nonautonomous systems it is relatively safe to take the use of a machine to signal that the operator has read the user manual and assumes responsibility for its use, except in cases where the machine fails to operate within the predefined limits (Matthias 2004). Thus, the user has control and is responsible for the actions and events that come from the normal operation of the system, but if it explodes or does something that was not stated in the manual when it should have been, we ought to blame the manufacturer. We know from the preceding discussion that, when machines capable of being ever so slightly autonomous are introduced, because of the rigidity and limited nature of nonautonomous systems, moral responsibility is complicated. Indeed, the agent responsible for operating this machine loses an element of control over the system. What happens if we progress further? Matthias (2004) argues that, if a NASA technician was operating a semiautonomous space vehicle and the vehicle falls into a crater between inputs because of long response times, we should not consider the technician responsible. Task-autonomous autonomous systems create a buffer between agent and system as well as another buffer between action and event, giving their operators the potential ability to increase their workload by operating multiple drones, but at the cost of understanding and situational awareness. The locus of responsibility categorically shifts away from the operator. However, the real problems arise when it comes to intelligent machines that are capable of adapting and learning new skills. That is, robotic systems for which unpredictability in their operations is a feature rather than a computer glitch or technical failure (Millar and Kerr 2012). Matthias (2004) asks us to imagine how we would impose moral responsibility if we were to revisit the space vehicle case and stipulate that it will not be remotely controlled from Earth, but rather have its own integrated navigation and control system, capable of storing data in its internal memory, forming representations and taking action from these. It should, therefore, be able to record video imagery and estimate the difficulty of crossing any familiar terrain. He asks, in this revised case, whom we should hold responsible if the vehicle were to once again fall into the crater?

Sparrow (2007) subsequently takes up this question in the context of his discussion of autonomous systems. He has us imagine that a drone, directed by sophisticated artificial intelligence, bombs a platoon of enemy soldiers who have indicated their intent to surrender. Who should we hold morally responsible for a particular event when the decision to bomb is made by an autonomous weapons system without a human operator? The reader's first intuition is probably to say that the responsibility for any moral crimes or just war violations rests with the developer of the weapon. However, Sparrow (2007) objects to this by relying on the user manual analogy. This analogy says that this would be unfair if it is a declared system limitation that the machine may attack the wrong targets in some percentage of cases. If this is the case, he suggests it may be the responsibility of the user (since the user is assumed to have read the manual). Sparrow (2007) also says that to hold the programmers or manufacturer responsible for the actions of their creation, once it is turned on and made autonomous, would be analogous to "holding parents responsible for the action of their children once they have left their care." Sparrow (2007) assumes this is wrong and that it naturally leads us to consider holding the commanding officer responsible. Yet again, he views this as unfair and thinks that to do so calls into question the nature of our "smart" weapons. If the machines start to make their own targeting decisions, he suggests that there will come a point at which we cannot hold the commanding officer responsible for any of the ensuing deaths (Sparrow 2007). It will be argued in the next section that Sparrow is mistaken about the assumed wrongness of sharing responsibility and that we are not yet at the tipping point he describes.

The final possible locus of responsibility under Sparrow's account, however, is the machine itself. Moral responsibility is typically attributed to moral agents; at least in the Anglo-American philosophical tradition, moral agency has been reserved for human beings. Unlike the majority of animals, rational human beings are seen as able to freely deliberate about the consequences of their actions and choose to act in one way or other, meaning that they are originators of *morally significant* actions. Although some people tend to anthropomorphize military robots and notable philosophers like Daniel Dennett (1997) and John Sullins (2006) have argued that they could be classed as moral agents, Sparrow (2007) argues that they should not and objects to the idea that they could have the kind of capacities that make human beings moral agents. He argues that it is unlikely that they will ever have the mental states, common sense, emotion, or expressivity equivalent to those of humans, and that, if they do develop these things, it would undermine the whole point of utilizing robots instead of human beings (Sparrow 2007). According to the argument, they would hold a moral status equivalent to that of human beings. But this is disputable, as it could be argued that, while artificial moral agents may be worthy of moral consideration, they would still hold a different status to biological moral agents by virtue of some natural/artificial distinction that gives greater weight to means of creation. Even if robots do not acquire human-level moral status, lower levels of machine autonomy may be sufficient for us to hold robots responsible. However, Sparrow (2007) holds that no robot can be held responsible because they cannot suffer. This presupposes that suffering is a requirement for responsibility, a presupposition not supported by the responsibility framework embedded in just war theory, but we will return to this point after having summarized the alleged problem.

For Sparrow, like Matthias before him, we have reached or are about to reach an important impasse. We already have many machines in development and a limited number of those, which are in use, are task-autonomous and can decide on a course of action in some limited scenarios without any human input. Going forward, all indications point to there being machines with rules for action that are not fixed by their manufacturers during the production process and that are open to be changed by the machine itself during its operation. That is, these machines will be capable of learning from their surroundings and experiences. Conventionally, there are several loci of responsibility for the actions of a machine, but both Matthias (2004) and Sparrow (2007) argue that these robots will bring about a class of actions for which nobody is responsible, because no individual or group has sufficient control of these systems. These cases constitute Matthias's "responsibility gap." At first blush, it might seem that this is basically the problem of many hands—the classical problem described earlier but with new relevance to the emergence of autonomous systems and the prospect of fully autonomous weapons systems. But to assume this would be mistaken. The argument advanced by Matthias and Sparrow is not that we cannot identify who is responsible, but simply that nobody is responsible. Sparrow would likely argue that, if there is any problem identifying the relevant persons, it is because they do not exist. In his article on corporate responsibility, Philip Petitt (2007) refers to the matter described as the "problem of no hands." This is a slight but important twist on Thompson's more familiar "problem of many hands," as described earlier and characterized by the widespread relinquishment of moral responsibility.

The problem that proponents of the responsibility gap put forward is fairly straightforward, though certainly not indisputable. However, to differentiate his argument from Matthias's, Sparrow (2007) suggests that we might better conceptualize his dilemma if we consider another case in warfare where the waters are somewhat muddied: the use of child soldiers. This analogy is outlined because it will be useful in problematizing Sparrow's argument, though the similarity between child and machine learning is not as great as Sparrow indicates. He says that like robots, one of the many reasons why it is unethical to utilize children in combat is that it places decisions about the use of force in the hands of agents that cannot be held responsible for them. According to him, child soldiers lack full moral autonomy, but they are clearly autonomous in some respect and "certainly much more autonomous than any existing robot" (Sparrow 2007). He goes on to say that, while they are not capable of understanding the full moral dimensions of what they do, they possess sufficient autonomy to ensure that those who order them into action do not or cannot control them, presenting problems for any effort to hold those who gives the orders exclusively personally responsible for the child soldiers' actions (Sparrow 2007). The idea Sparrow advances is that there is a conceptual space in which child soldiers and military robots are sufficiently autonomous to make the full attribution of responsibility to an adult or conventional moral agent problematic, but not autonomous enough to be held fully responsible themselves. Sparrow argues that his opponents try to close this space by stipulating that the relevant entities hold more or less responsibility than they should and thus fit within one of the polar boundaries, but that this does not adequately or fairly resolve the problem. He thinks that we should, in fact, ban the use of autonomous weapons altogether (Sparrow 2007). The next section will propose that we can actually handle this problem by moving to a more collective, pragmatic, and forward-looking notion of shared responsibility.

Toward a Revised Notion of Responsibility

Having outlined the importance of responsibility in the just war framework, explored some of the challenges that semiautonomous systems pose for responsibility attribution, and described the dilemma over responsibility for autonomous systems, the need for a revised notion of responsibility should be clear. But, to clarify, the arguments of both Matthias and Sparrow hinge on three basic premises. The first is that programmers, manufacturers, commanding officers, and the like may not be able to foresee what an autonomous robot, capable of learning, will do in the highly complex and dynamic military operating environment. The second is that either independently of or related to the fact that none of these agents are able to exert full control over the development or subsequent deployment of these systems, harm to others may eventually occur. The third is that an agent can only be held responsible for these harms if they have control in the sense that they have an awareness of the facts surrounding the action that leads to the harm and are able to freely manipulate the relevant causal chains based on these facts. The conclusion stipulates that since this is not the case as it pertains to programmers, manufacturers, or commanding officers, there is some sort of moral void created by the deployment of these systems, one that cannot be bridged by our traditional concepts of responsibility. While the problem is clear, it is not obvious that the overall conclusion can be accepted at face value or that any individual premise is correct. There are a number of points at which the alleged responsibility gap can be overcome, or, at least, a number of premises that can be called into question in order to cast doubt over the supposed insurmountability of the problem at hand.

In discussing the nature of this alleged responsibility gap and showing its inadequacy as a justification for a moratorium on autonomous systems, it is important to point out that the scope of the conditions for imposing responsibility has been overstretched or considered in too wide a frame. As will soon be shown with reference to the idea of shared responsibility, it is not impossible to impose responsibility in situations in which no individual has total control over the function of an autonomous system. Both Matthias and Sparrow go too far in suggesting that programmers, manufacturers, or commanding officers are freed from any form of responsibility because they do not have total control over the operation or manufacture of autonomous systems. An appeal to common sense should reveal that it is absurd and potentially dangerous to identify the failings of multiple individuals in their development and deployment and then deny their moral responsibility (Gotterbarn 2001). To do so is to deny an opportunity for the rectification of past, present, and future wrongs. As opposed to what might be stipulated by strict liability law, such a strong sense of control is not necessary for the imposition of some degree of moral responsibility. The relevant programmers, designers, manufacturers, and commanding officers are all responsible to some degree or extent.

Take Sparrow's claim that to hold the programmer of a dynamic learning machine responsible for its actions would be analogous to holding a parent responsible for the actions of their child once they are out of their care. While there are limits to any such analogy because of the varied learning mechanisms employed by child and machine, he seems to ignore the fact that parents are at least partially responsible for preparing their children for that moment when they leave their care and become independent, just as pharmaceutical companies are responsible for their pills once they have been manufactured and have entered the marketplace. In much the same way, the developers of autonomous systems hold significant responsibility for ensuring that their robots can operate as desired once given independence, something that is still a long way off in the majority of cases. Also, take the case of the commanding officer and continue with the parenting analogy. When a parent is teaching their child how to drive, for instance, the parent places the child in an area where the child can learn the necessary skills without risking her/his own safety or that of anyone else. Similarly, the commanding officer of an autonomous system has a responsibility to ensure that the system has been thoroughly prototype tested or is placed in an appropriate learning or test bed environment until such time that it performs at least as well as a manned system or until the chance of any serious harm occurring is so tiny that we can deal with it via the "functional morality" described later on, which recognizes that engineers and manufacturers will often choose to release intelligent machines with remnant unpredictability and that reprogramming for minor errors can assume the place of punishment.

The machine's path to full autonomy is a long (if not impossible) one, and Sparrow points to this using the child warrior analogy. Machines will not just "wake up," as is depicted in films about human-hating "terminators." Indeed, there is simply no way in which someone could deliberately create such an entity without a collective effort on the scale of the Manhattan Project. The general lesson to be drawn from this is that all the involved agents and any others associated with the use of autonomous systems (including the user in the case of semiautonomous systems) retain a share of responsibility, even though they may claim that they were not in complete or absolute control. It would be foolhardy, or even dangerous, to conclude from the observation that responsibility is obscured by the use of autonomous weaponry that nobody is, or ought to have been, held to account and that it is impossible to deal with the case of autonomous systems. On the contrary (and as others have argued in relation to informatics more generally) we are at such an important junction in the development of autonomous systems that we have reason to adjust and refine our conception of moral responsibility and to leave behind the idea that the imposition of moral responsibility relies on agents having full control over all aspects of the design or deployment of advanced robotics. This is because these systems are so complex that few of the design, development, or deployment related decisions are made on an individual basis. Why concentrate on the intentions and actions of humans alone in our moral evaluation of these systems when no human exerts full control over the relevant outcomes? We need to move away from the largely insufficient notion of individual responsibility, upon which we typically rely, and move toward a more complex notion of collective responsibility, which has the means and scope to include nonhuman action. That is, it must be a holistic approach that is capable of acknowledging the contribution of various human agents, systems, and organizations or institutions. The need to update our moral values and associated notions of responsibility will become more important as the technology develops, the risks increase, and the number of potential responsibility-related problems accumulate.

It is worth noting that others, foreseeing the difficulties that we are now facing with the development of things like intelligent autonomous robots, have already thought about calls for change in the way we think of responsibility. For instance, both Daniel Dennett (1973) and Peter Strawson (1974) have long held that we should conceive of moral responsibility as less of an individual duty and more of a role that is actively defined by pragmatic group norms. This argument in endorsed here, primarily because more classical accounts raise endless questions concerning free will and intentionality that cannot be easily resolved (if at all) from a practical perspective aimed at achieving results here and now. This sort of practical account has the benefit of allowing nonhuman entities, such as complex socio-technical systems and the corporations that manufacture them, to be answerable for the harms to which they often cause or contribute. It seems to require that we think in terms of a continuum of agency between nonmoral and full moral agents, with the sort of robots we are concerned with here falling just short of the latter. This pragmatic (or functional) approach also allows for the fact that agency develops over time and shifts the focus to the future appropriate behavior of complex systems, with moral responsibility being more a matter of rational and socially efficient policy that is largely outcomes-focused.

For our purposes here, it is useful to view moral responsibility and this pragmatic account in line with the social contract argument put forward much earlier. That is, we should view moral responsibility as a mechanism used by society to defend public spaces and maintain a state of relative harmony, generated under the contract by the power transferred from the individuals to the state. The end of this responsibility mechanism is, therefore, to prevent any further injury being done to society and to prevent others from committing similar offenses. While it might be useful to punish violators of the social contract and the just war theory that it has been argued follows from the social contract, it is not strictly necessary, nor is it necessarily the best approach to preventing harm in all scenarios. As Gert-Jan Lokhorst and Jeroen van den Hoven (2012) have argued, treatment is in many cases an equally effective option for the prevention of harm and one that we can apply to nonhuman agents in different forms, whether it is psychological counseling in humans or reengineering or reprogramming in the case of robots. This is important because it means that there is sufficient conceptual room between "operational morality and genuine moral agency" (Wallach and Allen 2009) to hold responsible, or, in Floridi and Sander's (2004) language, hold morally accountable, artificial agents that are able to perform some task and assess its outcome.

Scholars in the drone debate also seem to have become fixated on a backward-looking (retrospective) sense of responsibility, perhaps because even engineers and programmers have tended to adopt a malpractice model focused on the allocation of blame for harmful incidents. However, an effective and efficient responsibility mechanism that remedies the supposed gap should not only be about holding someone responsible only when something goes wrong. Therefore, this backward-looking sense of responsibility must be differentiated from forward-looking (prospective) responsibility, a notion that focuses more on capacity to effect change than blameworthiness or similar (Gotterbarn 2001). That is, at some point, we must stop thinking purely about past failures to take proper care and think about the reciprocal responsibility to take due care in future action. This is because in debates about real world problems such as the deployment of increasingly autonomous weaponry, we will also want our conception of responsibility to deal with *potential* problems. To this end, we can impose forward-looking or prospective responsibility to perform actions from now on, primarily in order to prevent undesirable consequences or to ensure a particular state of affairs obtains more effectively and efficiently than through the alternative means of backward-looking models. On this account, what matters is not past behaviors but a commitment to take actions and assume responsibilities that will enhance the ethical application of military power via autonomous weapon. It establishes a general obligation on the part of all those involved in the design, manufacture, and use of autonomous systems to give regard to future harms. Admittedly, as Seamus Miller (2008) has pointed out in the case of computing, it is difficult to reach any solid conclusion on how far into the future they are required to foresee. That said, two things are clear in the case of military robotics: first, if the recent troubles plaguing drone systems are any indicator, agents have reasonable grounds to expect the unexpected. Second, once fully autonomous systems are developed and deployed, no amount of policy making will stop their spread.

For the latter reason in particular, we have to think more carefully about where the majority of the forward-looking responsibility falls. In discussing the claim that wealthy countries must do more than comparatively poor countries to combat climate change and making use of Kant, James Garvey (2008) argues that in much the same way that "ought implies can," "can implies ought" in a range of other circumstances where the financial or political means behind the "can" have contributed to the problem that ought to be corrected or mitigated. This seems also to hold true in the responsibility debate with which we are engaged. Generally speaking, the more power agents have and the greater the resources at their disposal—whether intellectual, economic, or otherwise—the more obligated they are to take reasonable action when problems arise. Jessica Fahlquist (2008) has proposed a specific approach to identifying the extent of a person's obligation in relation to environmental protection based on varied levels of capacity to contribute to social causes, and given that power to enact change varies within the military and military–industrial complex to much the same extent as in the environmental world (and with some convergence), there is little reason why this should not be applied to the drone debate and extended to cover the manufacturers of autonomous systems as well as the governments that regulate them. The companies of the military–industrial complex are in a unique position, and have it well within in their power, to anticipate risks of harm and injury and theorize about the possible consequences of developing learning systems. The costs of doing so after the fact are great and many. Moreover, manufacturers are best positioned to create opportunities for engineers to do what is right without fear of reprimand, whether that would be going ahead as planned, designing in certain limitations in the system, or simply refusing to undertake certain projects. It therefore seems reasonable to impose forward-looking responsibility upon them. However, we know that profit can sometimes trump morality for these collective agents of the military–industrial complex, so concerned parties should also seek to share forward-looking responsibility and ascribe some degree of responsibility to the governments that oversee these manufacturers and set the regulatory framework for the development and deployment of their product, should manufacturers fail to self-regulate and establish appropriate industry standards.

Note that, while it may seem that the point advanced in this chapter is pitched against Sparrow's argument for a prohibition on the development and subsequent use of these systems, this is only partly true. The objection here is only to the antecedent of his claim or the vehicle that he uses to reach his final conclusion against the use of "killer drones." That is, it is not at all obvious that Sparrow needs to make the very bold claim that nobody can be held responsible for the use of military robots that fall between having operational autonomy and genuine moral autonomy. It seems that many other agents are sufficiently responsible and that, through embracing an instrumental approach toward backward-looking responsibility and combining it with a forward-looking account of responsibility put forward earlier, it is possible to distribute responsibility fairly. It may also be that some of the relevant agents, namely governments, reach the same conclusion as Sparrow. That is, while states and their militaries have a contractual obligation to effectively and efficiently protect the citizens who grant them power, it may turn out that this would mean avoiding the use of autonomous systems in some circumstances; indeed, there is some evidence of this in the implications and consequences described in earlier chapters. In fact, as unlikely as it seems, time may prove that autonomous systems pose such a problem that they warrant making a concerted effort to form an international treaty or a new Geneva Convention banning their use in all but very particular cases, perhaps where there is clearly demarcated "kill zone" in some remote part of the world. However, this is not yet clear and in light of the features of the revised account of responsibility advocated here, there is no intrinsic responsibility gap that warrants a prohibition of the use of autonomous systems, at least not without reference to the problems discussed in other chapters.

Critics are likely to argue that any account that allocates too much responsibility to organizational or governmental actors will erode the sense of personal responsibility that individuals feel and will not have the desired effect of improving the attainment of just outcomes with autonomous systems. What will happen, they may ask, if individual parties to collective atrocities are excused from the responsibilities that they would otherwise have under traditional notions of responsibility? There are a few ways to respond to this worry. The first consists in stressing that individuals are not in fact freed or excused from responsibility according to either the backward-looking or the forward-looking model. Individual agents will still be held causally responsible for their part in any war crimes under the functional/pragmatic account, so if a programmer or a commanding officer gets an entire village blown up by mistake, they will be held accountable. They will also be encouraged to ensure that systems are designed, developed, tested, and used in the desired way through the imposition of forward-looking responsibility. The second way to respond is to reaffirm that, because individual agents are the core units of the social contract, it would obviously be ideal if they were to gradually begin to embrace the right values and do the right thing, but allow that, in some circumstances, it is fairer or more effective to distribute the burden of action between individuals, institutions, and governments. This is due to the fact that, in the short to medium term, it must be recognized that both human and nonhuman agents will make mistakes that will lead to violations of some of the principles of the just war theory. As a consequence, the greatest share of responsibility must be ascribed to the most capable agents in the relevant scenario. In trying to highlight the oversimplified analysis of duties in rights-based theories, Henry Shue (1988) suggests that in some circumstances we must look beyond individuals and distribute responsibility to institutions in the most effective and efficient fashion relative to the time we have as well as the nature and severity of the problem.

For it is indeed institutions, such as judicial systems and police forces, upon which the duty to provide physical security ultimately falls. In the case of Sparrow's "killer robots," it seems best to impose much of the forward-looking responsibility on the government or its relevant standards departments, just because the government is best placed to ensure that systems are designed to rule out or reduce the impact of such mistakes or that measures are put in place to do the same thing.

Again, it must be stressed that this does not mean that individual agents or corporations with the relevant capabilities and resources are excused from efforts to achieve the desired effects or their share of responsibility, but merely that governmental agents ought to take greater efforts because the proposed account of responsibility can track their role and they can be held responsible for some violations of the just war theory.

Conclusion

This chapter began by exploring the need for a clear account of responsibility in just war theory and countering the claim that there is some sort of explicit requirement to hold a single individual responsible. There is, less specifically, an implicit responsibility component that stipulates that agents of war—whether human, nonhuman, or some combination thereof—must be held responsible for violations of the just war theory no matter how difficult the moral accounting. In the second section, it was demonstrated that technology generates a number of barriers to attribution of responsibility, from distancing users from their sense of responsibility to obscuring causal chains, making it more difficult to identify where a moral fault lies. The third section outlined how these issues and others come together in the case of fully autonomous warfare to create what Sparrow—following Matthias—alleges is a "responsibility gap," or a class of actions for which nobody/nothing is supposedly responsible. The final section laid the foundations for a theory of responsibility, which revolves around the idea that action and responsibility can be distributed among human and nonhuman agents or some combination thereof. More work is needed to reveal exactly what this new theory of responsibility will look like and to determine its precise implications, but if nothing else this chapter has hopefully demonstrated that, while ascribing responsibility in the case of autonomous systems is more complex and troubling than in the case of semi- and nonautonomous systems, it is by no means an insurmountable problem.

Notes

1. An earlier version of this chapter originally appeared in J. Galliott, *Military Robots: Mapping the Moral Landscape* (Surrey, UK: Ashgate, 2015).

References

Campbell, R. H., and G. Eaton. 2011. "It's All about You: Implementing Human-Centered Design." In *Human Performance Enhancement in High-Risk Environments: Insights, Developments, and Future Directions from Military Research*, edited by P. O'Connor and J. Cohn. Santa Barbara, CA: ABC-CLIO.

Coleman, S. 2012. *Military Ethics: An Introduction with Case Studies*. Oxford: Oxford University Press.

Cummings, M. L. 2004. "Automation Bias in Intelligent Time Critical Decision Support Systems." Paper presented at the AAIA First Intelligent Systems Technical Conference, Chicago.

Dennett, D. 1973. "Mechanism and Responsibility." In *Essays on Freedom of Action*, edited by T. Honderich. Boston: Routledge and Keegan Paul.

Dennett, D. 1997. "When HAL Kills, Who's to Blame? Computer Ethics." In *HAL's Legacy: 2001's Computer as a Dream and Reality*, edited by D. Stork, 351–65. Cambridge, MA: MIT Press.

Fahlquist, J. N. 2008. "Moral Responsibility for Environmental Problems—Individual or Institutional?" *Journal of Agricultural and Environmental Ethics* 22: 109–24.

Fieser, J., and B. Dowden. 2007. "Just War Theory." *Internet Encyclopedia of Philosophy*. http://www.iep.utm.edu/j/justwar.htm.

Fischer, J. M., and M. Ravizza, eds. 1993. *Perspectives on Moral Responsibility*, Ithaca, NY: Cornell University Press.

Fischer, J. M., and M. Ravizza. 1998. *Responsibility and Control: A Theory of Moral Responsibility*, Cambridge: Cambridge University Press.

Floridi, L., and J. Sanders. 2004. "The Foundationalist Debate in Computer Ethics." In *Readings in CyberEthics*, edited by R. Spinello and H. Tavani. Sudbury, MA: Jones and Bartlett.

Friedman, B. 1990. "Moral Responsibility and Computer Technology." Paper presented at the Annual Meeting of the American Educational Research Association, Boston.

Galliott, J. 2013a. "Unmanned Systems and War's End: Prospects for Lasting Peace." *Dynamique Internationales* 8: 1–24.

Galliott, J. 2013b. "Closing with Completeness: The Asymmetric Drone Warfare Debate." *Journal of Military Ethics* 11: 353–56.

Galliott, J. 2016. "Defending Australia in the Digital Age: Toward Full Spectrum Defence." *Defence Studies* 16: 157–75.

Galliott, J. 2017. "The Limits of Robotic Solutions to Human Challenges in the Land Domain." *Defence Studies* 17: 327–45.

Garvey, J. 2008. *The Ethics of Climate Change: Right and Wrong in a Warming World*. New York: Bloomsbury.

Gotterbarn, D. 2001. "Informatics and Professional Responsibility." *Science and Engineering Ethics* 7: 221–30.

Gray, C. S. 1997. "AI at War: The Aegis System in Combat." In *Directions and Implications of Advanced Computing*, edited by D. Schuler. New York: Ablex.

Grossman, D. 1995. *On Killing: The Psychological Cost of Learning to Kill in War and Society*. Boston: Little, Brown.

International Committee for Robot Arms Control. 2013. *Scientists' Call*. Accessed November 22, 2013. http://icrac.net/call/.

Jonas, H. 1984. *The Imperative of Responsibility: In Search of an Ethics for the Technological Age*. Chicago: University of Chicago Press.

Keller, B. 2013. "Smart Drones." *New York Times* [online], March 17. http://www.nytimes.com/2013/03/17/opinion/sunday/keller-smart-drones.html?pagewanted=all.

Lokhorst, G.-J., and J. Van Den Hoven. 2012. "Responsibility for Military Robots." In *Robot Ethics: The Ethical and Social Implications of Robotics*, edited by P. Lin, K. Abney, and G. Bekey. Cambridge, MA: MIT Press.

Manders-Huits, N. 2006. "Moral Responsibility and IT for Human Enhancement." Association for Computing Machinery Symposium on Applied Computing, Dijon.

Matthias, A. 2004. "The Responsibility Gap: Ascribing Responsibility for the Actions of Learning Automata." *Ethics and Information Technology* 6: 175–83.

McMahan, J. 2004. "The Ethics of Killing in War." *Ethics* 114: 693–733.

McMahan, J. 2009. *Killing in War*. Oxford: Oxford University Press.

Millar, J., and I. Kerr. 2012. "Delegation, Relinquishment and Responsibility: The Prospect of Expert Robots." Coral Gables, FL: University of Miami School of Law. http://robots.law.miami.edu/wp-content/uploads/2012/01/Kerr_Millar_Delegation-Relinquishment-Responsibility.pdf.

Miller, S. 2008. "Collective Responsibility and Information and Communication Technology." In *Information Technology and Moral Philosophy*, edited by J. Van Den Hoven and J. Weckert. Cambridge: Cambridge University Press.

Noorman, M. 2012. *Computing and Moral Responsibility*. http://plato.stanford.edu/archives/fall2012/entries/computing-responsibility/.

Petitt, P. 2007. "Responsibility Incorporated." *Ethics* 117: 171–201.

Rogers, W., and S. Rogers. 1992. *Storm Center: The USS* Vincennes *and Iran Air Flight 655*. Annapolis, MD: Naval Institute Press.

Royakkers, L., and R. van Est. 2010. "The Cubicle Warrior: The Marionette of Digitalized Warfare." *Ethics and Information Technology* 12: 289–96.

Shue, H. 1988. "Mediating Duties." *Ethics* 98: 687–704.

Sparrow, R. 2007. "Killer Robots." *Journal of Applied Philosophy* 24: 62–77.

Strawson, P. F. 1974. "Freedom and Resentment." In *Freedom and Resentment and Other Essays*. London: Methuen.

Sullins, J. P. 2006. "When Is a Robot a Moral Agent?" *International Review of Information Ethics* 6: 24–30.

Thompson, D. 1980. "Moral Responsibility and Public Officials: The Problem of Many Hands." *American Political Science Review* 74: 905–16.

Thompson, D. 1987. *Political Ethics and Public Office*. Cambridge, MA: Harvard University Press.

United States Department of Defense. 2009. "FY2009–2034: Unmanned Systems Integrated Roadmap." 2nd ed. Washington, DC: Department of Defense.

Waelbers, K. 2009. "Technological Delegation: Responsibility for the Unintended." *Science and Engineering Ethics* 15: 51–68.

Wallach, W., and C. Allen. 2009. *Moral Machines: Teaching Robots Right from Wrong*. Oxford: Oxford University Press.

Walzer, M. 2006. *Just and Unjust Wars: A Moral Argument with Historical Illustrations*. New York: Basic Books.

Zuboff, S. 1985. Automate/Informate: The Two Faces of Intelligent Technology. *Organizational Dynamics* 14: 5–18.

Chapter 7

Ethical Weapons

A Case for AI in Weapons

JASON B. SCHOLZ, DALE A. LAMBERT,
ROBERT S. BOLIA, AND
JAI GALLIOTT

Significant recent progress in AI is positively affecting everyday tasks, as well as science, medicine, agriculture, security, finance, law, games, and even creative artistic expression. Nevertheless, some contend that, on ethical grounds, military operations should be immune from the progress of automation and artificial intelligence evident in other areas of society. Human Rights Watch (IHRC 2012) has warned of "killer robots—fully autonomous weapons," believing that "such revolutionary weapons would not be consistent with international humanitarian law and would increase the risk of death or injury to civilians during armed conflict" and noted that the "primary concern of Human Rights Watch and IHRC [International Human Rights Clinic] is the impact fully autonomous weapons would have on the protection of civilians." The Campaign to Stop Killer Robots[1] echoes this sentiment, with signatories that include over a thousand experts in AI, as well as science and technology leaders such as Elon Musk, Steve Wozniak, Noam Chomsky, Skype cofounder Jaan Tallinn, and Google DeepMind cofounder Demis Hassabis. They define "The Problem" on their website[2] to include "allowing *life* or death decisions to be made by machines" (our emphasis), how machines "lack human judgement and the ability to understand context," and how such

The views are solely those of the authors and do not necessarily reflect the views or policies of the Australian Government or the Department of Defence.

systems create an "accountability gap as there is no clarity on who would be legally responsible" for their design, development, and employment.

With such compelling arguments from these luminaries, how could one disagree? We respond with a case for ethical weapons, explaining why a blanket prohibition on AI in weapons is a bad idea, and why "life" decisions should at times be made by machines. As noted by Lambert and Scholz (2005), automobiles rival wars as a contributor to human death, and yet the automobile industry is one of the leaders in integrating automated decision-makers into vehicles. The manufacturers' motivation is to make automobiles safer. We advocate a similar motivation in a military context.

A simple example serves to illustrate this point. Consider the capability of a weapon to recognize the presence of an international protection symbol—perhaps a Red Cross, Red Crescent, or Red Crystal—in a defined target area and abort an otherwise unrestrained human-ordered attack. Given the significant advances in visual machine learning over the last decade, such recognition systems are technically feasible.

So, inspired by vehicle automation, what we call an ethical weapon is a weapon with built-in safety enhancements. We further develop this safety argument for weapons by adapting the guidelines for ethics in autonomous vehicles developed in Germany.

An ethical weapon takes an attack order as input and makes a decision *not to obey* the order if it assesses the presence of unexpected *protected* object(s).[3] What we mean by protected may include legal-identified entities from Red Cross–marked objects through to persons hors de combat and policy-identified entities specified in rules of engagement. We recognize that ends of this spectrum range from easy to very difficult technological challenges for AI (Sparrow 2016). What this does mean is that some progress toward ethical weapons can be made immediately, and clearly any progress would constitute a humanitarian enhancement. For this reason, we examine the technological feasibility of these systems as complexity and uncertainty increase.

Lambert (1999) termed weapon systems with these ethical improvements *moral weapons* and included among the enhancements "fully integrated human-machine decision making," the option of "allowing the machine to at times override the human," and the ability to assess and *decline* targeting requests when rules of engagement violations are deduced, with the decisions to override these weapons logged for subsequent accountability review. We term these *ethical weapons*, rather than moral weapons, to avoid any potential confusion with moral responsibility, as we do not mean to imply that such weapons possess *moral responsibility*.

We assert that autonomy in weapons is not likely to be banned regardless of campaign efforts and advocate that critics, or those who generally reject the concept of autonomous weapons, might consider these new ethical guidelines to further reduce casualties over current weapons and address their central concern about humans losing control over decision-making in warfare.

A Case for Ethical Weapons

The world has placed prohibitions on the possession and use of certain types of weapons, including chemical, biological, nuclear, and potentially persistent unexploded ordnance such as cluster munitions and landmines. Prohibition of these weapons has not prevented states or nonstate actors from developing them. India, Pakistan, Israel, and North Korea have developed nuclear weapons, and Iran was actively developing a nuclear weapons program until 2009 (Barton 2015). Moreover, none of the nations that possess nuclear weapons are signatories to the Treaty on the Prohibition of Nuclear Weapons (White and Paterson 2017). Nevertheless, preventing nations or nonstate actors from acquiring nuclear weapons has been reasonably effective until now, but only because it has been possible to physically control access to the relatively difficult-to-obtain materials required to produce them. In the case of autonomous weapons, it is not the materials that are lacking, but the code. The development of the algorithms needed for autonomous weapons are in many cases the same as those needed for autonomous cars or mobile phone apps. It is not possible to identify certain types of code that are militarily useful and ban them. The construction of autonomous weapons once the component technologies—many of which will be in the public domain—become available is only a matter of time, and not only for nation-states. This is an area in which those states charged with maintaining international order do not want to find themselves lagging behind.

A blanket prohibition on "AI in weapons" would have unintended consequences, due to its lack of *nuance*. There is a distinction to be made about those *kinds of AI* that would have humanitarian benefits. This lack of nuance is also evident in the case against chemical weapons. For example, pepper spray or tear gas is a chemical agent banned in warfare under the Chemical Weapons Convention of 1993, making it illegal for use by militaries except in law enforcement. The denial of tear gas to military forces removes a less-than-lethal option from the inventory, which could lead to the unnecessary use of lethal force. Another consequence of a ban would be to deny the use of autonomous weapons as a countermeasure against other autonomous weapons.

The world has large stockpiles of weapons—bombs, mines, bullets, guns, grenades, mortars, and missiles—that have no inbuilt technical controls related to the conditions under which they are employed. This is perhaps a far more frightening reality of immediate humanitarian concern than any fictional scenario involving "killer robots." Munitions developed for use by militaries and the public generally possess no inbuilt safeguards that prevent them from being used by unauthorized persons. We must remember that military forces that cannot afford precision weapons are regularly legally justified in the defense of their state to kill enemy combatants with firearms, bombs, and other, sometimes imprecise and indiscriminate weapons. Yet as military technology becomes increasingly capable of yielding more precise outcomes at lower cost, and of halting an enemy without causing unnecessary suffering or harm to those nearby, this is a situation that moral philosophers and international law might now reconsider.

The concept of ethical weapons is relatively new, as the technology to enable them is new. One example is "smart guns" that remain locked unless held by an authorized user via biometric or token technologies to curtail accidental firings and cases of a stolen gun being used immediately to shoot people. These technologies might also record events including the time and location of every shot fired, providing some accountability. Another example is the 2004 US initiative to only use landmines with a self-neutralization ability to prevent persistence beyond the operational period.

Ethical weapons should be subject to a process of weapons review according to the provisions of Article 36 of Additional Protocol I of the Geneva Convention. A number of legal officers have already begun to give thought to how this might be accomplished for autonomous weapons (Backstrom and Henderson 2012; Copeland and Reynoldson 2017).

A Code for Ethical Weapons

In 2013 the Human Rights Council of the United Nations General Assembly made the recommendation that developers of lethal autonomous robots (LARs) "establish a code or codes of conduct, ethics and/or practice defining responsible behaviour with respect to LARs" (Human Rights Council 2013). As a starting point, one might look for similar codes in related fields. In July 2016 the German federal minister of transport and digital infrastructure (Bundesministerium für Verkehr und digitale Infrastruktur, BMVI) appointed an expert panel of scientists and legal experts to serve as a national ethics committee for autonomous vehicles (Luetge 2017). A year later they made headlines when they issued "the world's first ethical guidelines for driverless cars" (Tuffley 2017; BMVI 2017). Obviously, automobiles are not designed to be weapons, though their kinetic energy and ubiquity make them at least as deadly in practice, such that their automation raises a number of issues in terms of potential damage to life and property. Many of the normative questions that arise as a result, and the normative frameworks utilized to answer said questions, are similar. As such, it does not seem unreasonable to take the BMVI ethics code as a basis for the development of an analogous code for ethical weapons. This may be further justified after consideration of some of the relevant principles of the Law of Armed Conflict:

Military Necessity—for military operations, any use of weapons requires said use to produce military gains that are not otherwise prohibited by international humanitarian law (International Committee of the Red Cross 1949: 1868 St Petersburg Convention). The BMVI code does not address this issue, since the presumption is that automobiles have a right to be on the road for purposes of transport regardless of what or whom they are transporting, and thus needs to be augmented as part of the law of armed conflict.

Distinction—the ability to distinguish between the civilian population and combatants, and between civilian objects and military objectives, and accordingly direct operations only against military objectives (International Committee of the Red Cross 1949: Additional Protocol 1, Article 48). In the case of ethical weapons, they might identify protected symbols, noncombatants, surrendering persons, and persons who are hors de combat in order accordingly as (and if) the AI technologies continue to advance. Distinction is not used in the German automobile ethics code, except in the priority for human persons over nonhuman persons (i.e., animals) in the case of an impending accident. This, again, is included under the law of armed conflict.

Proportionality—An attack shall not be launched if it may be expected to cause collateral casualties or damage which would be excessive in relation to the concrete and direct military advantage anticipated from the attack as a whole (Doswald-Beck 1994). The analogy for automobiles has been the frequent subject of trolley problem studies (Bonnefon, Shariff, and Rahwan 2016) and is considered in the German ethics guidelines. But how much of an obligation do military strategists have to avoid harm to civilian populations? Customary international humanitarian law provides further useful protections beyond merely justifying proportionality on the basis of the principle of double effect, including rule 15 (precautions in attack); rule 20 (advance warning); and rule 24 (removal of civilians and civilian objects), which are applied in the following to ethical weapons.

The German ethics code opens with general remarks and a mission statement. We have adapted this as follows.

Ethical Weapons: Mission

Important decisions will have to be made concerning the extent to which the use of ethical weapons is required. States have a record of failing to intervene with new weapons technologies, even when doing so would have been justified. The character of the justification to employ ethical weapons could be understood in three ways. First, states with the capability and capacity to do so may be obliged to deploy ethical weapons and hence face blame should they decide otherwise. An argument for these capabilities potentially being obligatory is that ethical weapons improve humanitarian outcomes (e.g., reducing accidental deaths) without affecting military effectiveness and are likely to utilize technologies that are low cost due to their commercial scale, with further justification explained by Galliott (2015). Second, the development and deployment of ethical weapons could be supererogatory in the sense that it would be good for a state to intervene with ethical weapons in particular circumstances, but not ethically required.[4] Third, such action could be justified but be neither obligatory nor supererogatory, such that the use of ethical weapons would be ethically acceptable but likely to yield little benefit over the status quo. We suggest that in all cases where the use of ethical weapons is justified, that is, in the pursuit of just causes, their use is either ethically obligatory or supererogatory, but much hinges on the conditions in which they are used and the way in which they are designed. At a fundamental level, it comes down to a question: How much dependence on technologically complex systems—based on artificial intelligence and machine learning—are we willing to accept in order to achieve, in return, more safety for noncombatants, more safety for our military, who, acting on behalf of our society, warrant protection, better compliance with laws of armed conflict, and improved operational efficiency to defeat ever improving adversary capabilities? What precautions are needed to ensure appropriate competency, authority, and responsibility? What technological development guidelines are required to ensure that we do not blur the contours of a human society that places trust in its military commanders and their freedom of action, physical and intellectual integrity, and entitlement to social respect at the heart of its legal regime?

Ethical Guidelines

1. Purpose

The primary purpose of ethical weapons is to improve the safety of protected entities and noncombatants under the law of armed conflict and the rules of engagement. A secondary purpose is to increase freedom of maneuver for military commanders, thereby enabling further ethical benefits.

2. Positive Balance of Risks

The objective is to reduce the level of harm within the laws of armed conflict with the ultimate goal of zero unintended noncombatant casualties. The "fog of war" means that noncombatant casualties will be a reality in twenty-first-century warfare, but to minimize these should be the ultimate aim, made possible only by increasing the intelligence of weapons and effectors of all kinds. The adoption of ethical weapons is justifiable if it promises to produce a diminution in harm to human and political capital in comparison to conventional weapons.

3. Avoidance of Ethical Dilemmas to the Extent Possible

Ethical weapons should prevent noncombatant harm within the law of armed conflict wherever this is practically possible. Further, appropriate reduction in operator involvement might reduce risk of post-traumatic stress disorder. Based on the state of the art, the technology should be designed in such a way that critical situations do not arise in the first place. These include dilemma situations, in which ethical weapons or military commanders, or both, have to decide which of two "evils" to perform. In this context, the entire spectrum of technological options should be used and continuously evolved; for example, limiting the scope to certain controllable conditions in military environments, allowing the weapon to dynamically and cognitively choose a payload yield reduction below a maximum level authorized, making the payload inert, performing weapon avoidance maneuvers, producing signals or advance warnings for persons at risk, or deferring weapons to alternate targets of opportunity in time and space. The significant enhancement of noncombatant safety is the objective of development and regulation, starting with design and programming of the ethical weapons such that they track in a defensive and anticipatory manner, posing as little risk as possible to vulnerable noncombatants while still achieving their missions.

4. Armed Conflict Shall Be Managed by Mixed-Initiative Agreements

A statutorily imposed obligation to use ethical weapons is ethically questionable if it entails submission of *all* military commanders to technological imperatives. That is, there should be a prohibition on degrading humans to *only* being subservient elements in an autonomous network. Dynamic and recorded mixed-initiative agreements between humans and machines shall subsume hierarchical human-only command arrangements for ethical weapons.

5. Primacy of Human Life

In situations that prove to be unavoidable, despite all technological precautions being taken, protection of humans enjoys priority in a balancing of interests compared with damage to animals or property.

6. Military Commanders Decide to Sacrifice Specific Lives

Ethical weapons execute targeting according to processes approved by military commanders in compliance with applicable laws and rules of engagement.

7. Machines Minimize Innocent Casualties

In the event of situations where the death of innocent people is unavoidable, ethical weapons shall seek to minimize casualties among innocent people.

8. Military Commanders, Developers, and Defense Departments Are Accountable for Ethical Weapons

Military commanders throughout the network of command remain accountable for the use of ethical weapons. All ethical weapon systems will log the protocol exchange between ethical weapons and military personnel, as well as critical weapon status and knowledge, to provide accountability and postaction review from the perspectives of accountability of commanders, developers, and defense departments as a whole.

9. Security of Ethical Weapons

Ethical weapons are justifiable only to the extent that conceivable attacks, in particular manipulation of the information technologies they rely upon or other innate system weaknesses, do not result in such harm as to undermine confidence in the military or in ethical weapons.

10. Awareness and Recording of Responsibility Transfers

It must be possible to clearly distinguish whether an ethical weapon is being used where accountability lies and comes with the option of overruling the system. The human–machine interface must be designed such that it is clearly regulated and apparent where authority, competency, and responsibility lies, especially the responsibility for control. The distribution of responsibilities (and thus of accountability), for instance with regard to the time and access arrangements, should be reliably recorded and stored. This applies especially to human-to-technology handover procedures.

11. Human on- and off-the-Loop

The software and technology in ethical weapons must be designed such that the need for an abrupt handover of control to military commanders is minimized. To enable efficient, reliable, and secure human–machine communication and prevent overload, the systems shall adapt to human communicative behavior where possible, rather than requiring humans to enhance their adaptive capabilities. Communication to the human will be appropriately abstracted and sufficiently timely where feasible, noting that human-in-the-loop will give way to human-on-the-loop and human-off-the-loop relationships for periods of time.

12. Machine Self-Learning Considerations

Learning systems that are self-learning in training, operation, and in their connection to scenario databases may be allowed if, and to the extent that, they generate safety gains. Self-learning systems must not be deployed unless they meet the safety requirements for ethical weapons and do not undermine these guidelines.

13. Fail-Safe Management

In situations where protected marked objects or unanticipated noncombatants are present, the ethical weapon must autonomously (i.e., without direct human intervention) enter into a "safe condition." Identification of what constitutes safe conditions for weapon disposal and recovery, planning, and handover routines is required prior to ethical weapons use. This may include means under control of the machine to place the weapon in a location that has minimal human impact; neutralize explosives in the weapon by use of separated chemical components in warhead design that are diffused to prevent future ignition or exploitation; and reduce weapon kinetic energy and damage.

14. Military Education and Training

The proper use of ethical weapons should form part of military commanders' general education. The proper handling of ethical weapons should be taught in an appropriate manner during training, and teams of commanders and ethical weapons should be tested for capability certification.

The Feasibility of Ethical Weapons

How might these ethical guidelines be achieved using AI in ethical weapons?

Ethical weapon decision-making needs to be considered in terms of its integration with military decision-making or command and control (Lambert and Scholz 2005). By command, we mean the "creative expression of *intent* to another given *awareness*." By control we mean the "expression of a *capability* (a plan is an example of a capability) to another and the monitoring and correction of the execution of that capability given *awareness*." So, command is separable from control yet remains coupled by awareness. McCann and Pigeau (1999) categorize three dimensions to command: competency, authority, and responsibility.

If the reader can entertain that command may be separable from being human, then these combined notions help tease out the role of human and machine in the enterprise rather than lumping everything under "human command," which also serves to succinctly structure a response to critics' key concerns from earlier (IHRC 2012).[5] Through formal definition of the semantic meaning of these terms (to follow) we shall make explicit these notions so that they may be encoded into the control of weapons. This addresses a lack of "*meaningful* human control" that has been the central criticism of advocates for a ban on autonomous weapons. It also advances a complex discussion about the future roles of humans and machines in the military enterprise. By exercising some competency, authority, and responsibility, an ethical weapon could potentially ensure "the protection of civilians during times of war" (IHRC 2012) even without full "human judgment and the ability to understand context."

A framework for considering shared decision making is illustrated in figure 7.1. The achievement of intent using capability, given awareness,

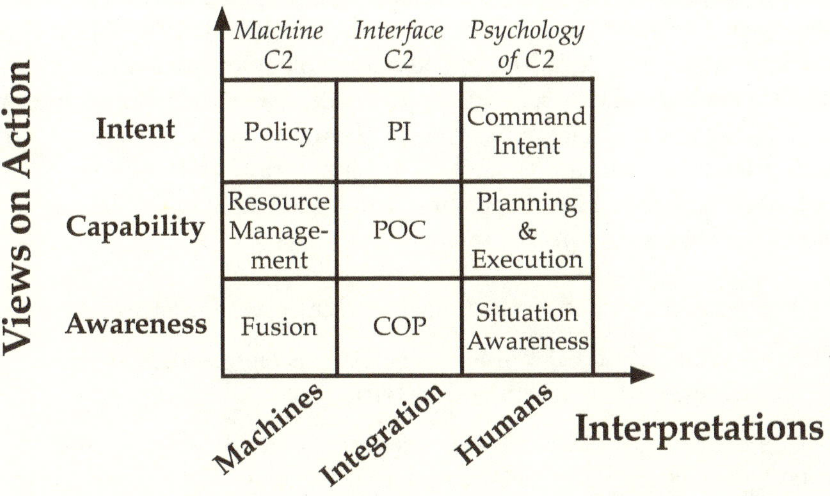

Figure 7.1. A model of shared command and control between human and machine from Scholz et al. (2012). COP = common operating picture, POC = planning and operating controls, and PI = policy interface.

occurs across the entire human and machine enterprise. The integration of humans and machines includes interfaces and interactions for managing policies, planning, operations and "common" (meaning agreed) pictures. As already illustrated, even simple *computer* vision capabilities, which are a form of improved awareness in the machine, embodied in a weapon seeker to identify a protected asset may offer significant immediate ethical gains. Our ambitions, however, are far greater. The full potential for ethical weapons rests on the degree to which AI can be embedded in the weapon system, subject to the ethical guidelines, to realize a spectrum of "human qualities" in machine terms. This chapter makes no presumptions about whether AI will realize the full spectrum of "human qualities," as the merits of ethical weapons require only some of those qualities. The most critical of these qualities correspond to command. In the following, we consider the research and technological feasibility of engineering and training machines as ethical weapons in terms of the elements of competency, authority, and responsibility.

Competency

Competency requires our ethical machines to be able to *act* by using their *capability* to achieve *intent* given *awareness* in an *emotionally* acceptable way. Each aspect is described in the sections below. While the machine might not replicate the neurophysiology of human cognition, it is nonetheless able to provide much of the *functionality* of cognition through alternate means, just as a jet engine attached to fixed wings with ailerons is able to accomplish the lift, thrust, and drag functions of flying without employing flapping wings. For example, in the case of emotions, we have cognitive software that

- recognizes emotions in human speech, text, and faces
- reasons while using internal emotional representations for happy, sad, disgusted, and so forth
- through a virtual adviser interface, responds emotionally through Ekman's facial emotions

These are important attributes for maintaining good human–machine relationships. Explicit implementations of ethical agents or "cognitive systems" are required (Scheutz 2017) and might be achieved using a variety of approaches (Arkin 2009) that include encoding legal theories; models of human-like moral competence; or via operationalizing ethical theories such as virtue ethics, deontology, or consequentialism.[6] Vanderelst and Winfield (2018) note several instances of robots equipped with limited ethical principles, in addition to their own architecture based on a simulation model of cognition. In contrast, most architectures are rule- or logic-based symbol manipulators. The ATTITUDE cognitive architecture developed by Lambert (2003) is *one* example of the latter that is both implemented and can integrate all of these competency elements.[7] We use the ATTITUDE system as a vehicle to demonstrate the feasibility of ethical weapons. Figure 7.2 provides a pictorial overview that is addressed in each of the following sections.

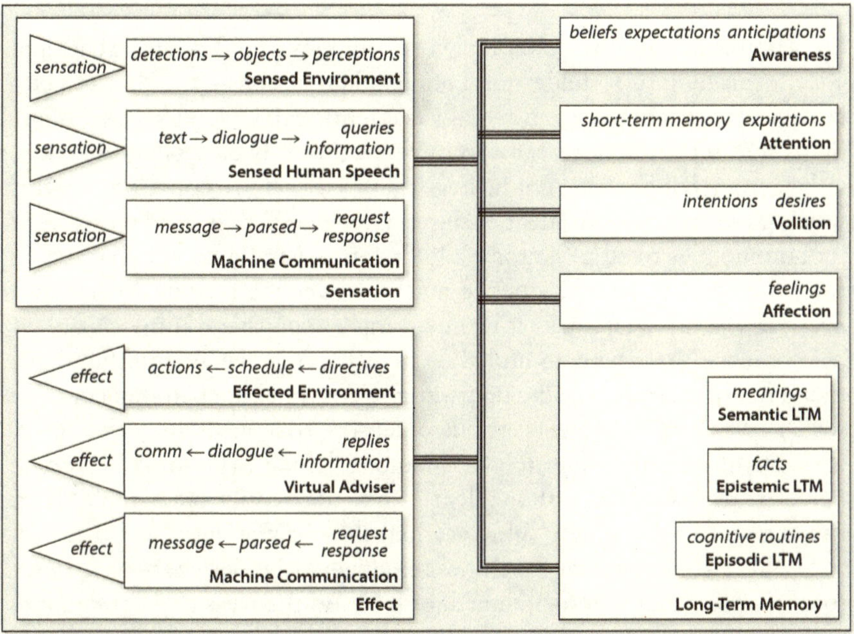

Figure 7.2. The ATTITUDE Cognitive Architecture provides an integrated implementation of all components of cognitive competency.

1.1.1 Intent

Daniel Dennett's (1971) intentional stance explains that humans are able to make some sense of the behaviors of so many complex things by treating them as agents. An agent is something that autonomously senses, thinks, and acts in the world. The intentional stance is not a description of the internal mechanisms of cognition and so remains neutral on details, allowing it to play a useful role as "a middle-level specifier of subsystem competencies (sub-personal agents, in effect) in advance of detailed knowledge of how they in turn are implemented."

> By exploiting this deep similarity between the simplest—one might as well say most mindless—intentional systems and the most complex (ourselves), the intentional stance also provides a relatively neutral perspective from which to investigate the differences between our minds and simpler minds. (Dennett 2009, 9)

So, whether it's a simple or complex system, positing it as an intentional system can help us to understand and explain actions—whether *or not* a biological or machine system actually is intentional. Ubiquitous road traffic lights that substitute for what was once, in the early days of automobiles, a dangerous police occupation, might also be usefully conceived in this way. A red light is, in effect, issuing a *command* as it expresses intent for automobiles to stop. Automobile drivers generally accept that intent (and thereby the authority of the machine over the human driver—as a result of the law that backs it up) by using a cognitive routine capability of *controlling* their vehicle and thereby achieving the intent they have individually accepted. Traffic lights in modern times are part of a complex road network of multiple distributed sensors that measure traffic flows, and thus it may be useful for autonomous vehicles to "think" of traffic lights in this way, too—after all, it is not necessarily predictable, from the perspective of a single vehicle, when these traffic lights will turn red.

We also assert that machines, engineered by designers to achieve some purpose, may be programmed, or learn on the basis of programming, without external supervision from experience, so as to expose purpose explicitly as intentions. This makes those designs more adaptable by the people who are to use them. This perspective is supported by Dennett (1971), who asserts that a functional or "design stance" is used by humans when the physical stance (based on physics) is too difficult to explain something, and further that the "intentional stance" is used when a functional description is inadequate.

Intentional systems theory takes a step beyond the intentional stance, motivating a desire to embed intentional representations within machines:

> Intentional systems theory is in the first place an analysis of the meanings of such everyday "mentalistic" terms as "believe," "desire," "expect," "decide," and "intend," the terms of "folk psychology" that we use to interpret, explain, and predict the behavior of other human beings, animals, some artifacts such as robots and computers, and indeed ourselves. (Dennett 1971, 1)

The term *cognitive machine* may be used to express the realization of computational processes involving *volitions* represented by verbs such as "desires" and "intends," *interactions* such as "perceives" and "informs," and *awareness* such as "believes" and "expects" in machines. Samsonovich (2010) includes ACT-R and JACK as examples of this approach to cognitive architectures. "Mentalistic" language is meaningful to humans and thus, if embedded to control machine operation, allows for a communicative connection between human and machine, and the expression of intentionality to another. Intentions may of course cascade throughout a social system, not only decomposed top-down in a traditional hierarchy, as in a network structure where intentions may flow, decompose, and be fused together (Lambert and Scholz 2005).

Machines have also demonstrated the ability to autonomously choose appropriate goals in adapting to unexpected changes in their environment (Jaidee et al. 2011) and to create new goals (Vattam et al. 2013). A machine that observes a discrepancy preventing it from achieving its goals may readily generate a new goal by reasoning over its explanation. This allows goals to be created within a given semantic framework, and, if augmented with techniques like zero-shot machine learning (Teney and Hengel 2016), offers the potential to create newly defined semantic representations never seen before by the machine, in order to further extend the scope of new goal creation.

Theory of mind will be an important capability for machines of the future in order to attribute mental states to themselves and to others, to understand that others have beliefs, desires, intentions, and perspectives that are different from their own. Social planning involving human and machine agents illustrates one computationally feasible approach to this based on modal logics (Miller et al. 2017).

Capability

Capability involves all three long-term memory elements for episodic (routines), semantic (meaning), and epistemic (facts) representation from figure 7.2. Kahneman (2011) has reflected that much human thinking is automatic—what he calls "System 1" or "thinking fast," as opposed to "System 2" or "thinking slow." Both types of thinking happen simultaneously and are bound by a limited attention budget. Thinking fast is characterized as *automatic*, frequent, subconscious, parallel, and emotional, whereas thinking slow is effortful, infrequent, conscious, sequential, and logical. In learning a new competency, such as driving a car, much "System 2" effort is required initially to learn subcompetencies, yet over time these become routine. Repetitive frequent training hones competence and, in the process, those routine actions become embedded as cognitive routines,[8] which are thereafter treated as primitive. Thus, for an experienced driver, driving is most of the time automatic.

However, cognitive routines for automation should only be applied under strict invariant or controlled conditions, "context" or "situations." Variation of critical conditions when operating a fixed routine demonstrates the fragility of nonadaptive automation. Driving your own car on a familiar highway in fine weather and light traffic invokes a cognitive routine, but as soon as a driver unexpectedly pulls into the road in front of you, the routine is interrupted, and for those trained in defensive driving, the new situation may be familiar enough to invoke another cognitive routine for crash avoidance without losing control of their own car.

Military operations are often characterized by extreme physical and mental conditions, exacerbated by ubiquitous uncertainty. Armed forces cope with these situations by means of frequent, specialized, realistic training exercises to expose personnel to conditions analogous to those under which they will have to operate in combat. Yet not everything that might happen can be trained for, and fundamental uncertainty remains the rule rather than the exception in military operations. The training is designed not merely to allow soldiers to act according to a specific routine, but to recognize similarities and differences between their training scenarios and the situations in which they find themselves and adapt accordingly. Humans not only acquire cognitive routines through training but also comprehend the situations in which these routines apply, whether explicitly or implicitly, and adapt or swap out routines to suit. Thus, autonomy will not be achieved by cognitive automation alone, but will require contextual adaptation of that automation. Automation provides competencies with which to act. Autonomy provides competencies for managing changes in routines. In Attitude, a cognitive routine often performs both of these functions simultaneously because (a) cognitive routines progress with a success–fail construct so the routine can have alternative options within it; and (b) cognitive routines operate concurrently at multiple levels of abstraction, with a routine running with both superior and subordinate routines.

As noted by Sparrow (2016), "the most fundamental barrier to building an ethical robot is that ethics is a realm of meanings." Bolia and Slyh (2011) also identified the importance, and relative scarcity of, semantic reasoning capabilities for complex cognitive tasks encoded in machines. In order to formally describe concepts and embed those concepts in machines, Lambert and Nowak (2008) proposed a comprehensive semantic system called "Mephisto" that extends Dennett's physical, design (functional), and cognitive (intentional) levels by adding a social level above the intentional level and a metaphysical level below the physical level. This extends the notion of intentional systems theory to five levels. Examples of relevant concepts are shown in figure 7.3.

Social:	group, ally, enemy, neutral, own, possess, agree, invite, offer, accept, reject, authorize, allow.
Intentional:	cognitive, routine, learned, achieve, perform, succeed, fail, intend, desire, belief, expect, anticipate, effect, affect, approve, disapprove, prefer.
Functional:	sense, move, attach, inform, operational, attack, disrupt, neutralize, destroy, can.
Physical:	land, sea, air, outer space, incline, decline, number, temperature, weight, energy.
Metaphysical:	exist, fragment, identity, time, space, before, connect, distance, area, volume, angle.

Figure 7.3. Illustrative symbolic primitives and symbolic relations for machines, from Lambert and Nowak (2008).

It is Lambert's conjecture that semantics are "hardwired" in humans. As supporting evidence, Goddard (2010) offers a Natural Semantic Metalanguage that empirically examined a range of human languages and found a common set of around sixty semantic concepts that he conjectures form the basis of *all human language*. Rather than learn these semantic concepts, it is proposed that we learn *with them* as we apply the hard-wired concepts to our environment to form knowledge and belief. These primitive concepts may also be combined to form compound relations.

Combining meanings (semantic) with routines (episodic) and facts (epistemic), it is possible to formally define competency for individual X to achieve outcome α, such that, X being a cognitive individual, if there exists ($\exists$) a routine R that X can perform resulting in a behaviour ξ, and whenever R is performed by X, the behaviour ξ achieves outcome α:

$$\text{competency}(X, \alpha) = \text{cognitive}(X) \mathbin{\&} \exists R \, \exists \xi \, (\text{routine}(R) \mathbin{\&} \text{can\_perform}(X, R) \mathbin{\&} (\text{performs}(X, R, \xi) \Rightarrow \text{achieves}(\xi, \alpha)))$$

Thus, automation routines and their adaptation form the foundation of "capability" described earlier, recounting that the goal of command and control is the achievement of intent using capability, given awareness.

1.1.2 Awareness

Awareness underpins command and control for the achievement of intent using capability (i.e., routines and their adaptation). Endsley (1988) defines human situation awareness as "the perception of the elements in the environment within a volume of time and space, the comprehension of their meaning, and the projection of their status in the near future." For machine processing, the widely adopted (and adapted) Joint Directors of Laboratories model for data fusion (Hall and Llinas 1997) provides several core levels illustrated in figure 7.4.

Lambert (2001) noted that information fusion may be regarded as situation awareness performed by machines. In this case perception aligns with object refinement assessment, comprehension with situation assessment, and projection with scenario assessment.

Object assessment combines sensor-derived information and is based on numerical measurements. Combining the "sensations" measurements from various complementary sensors provides a reliable estimate irrespective of environmental conditions, including obscuration due to weather, smoke, or other conditions. Object assessment includes detection and

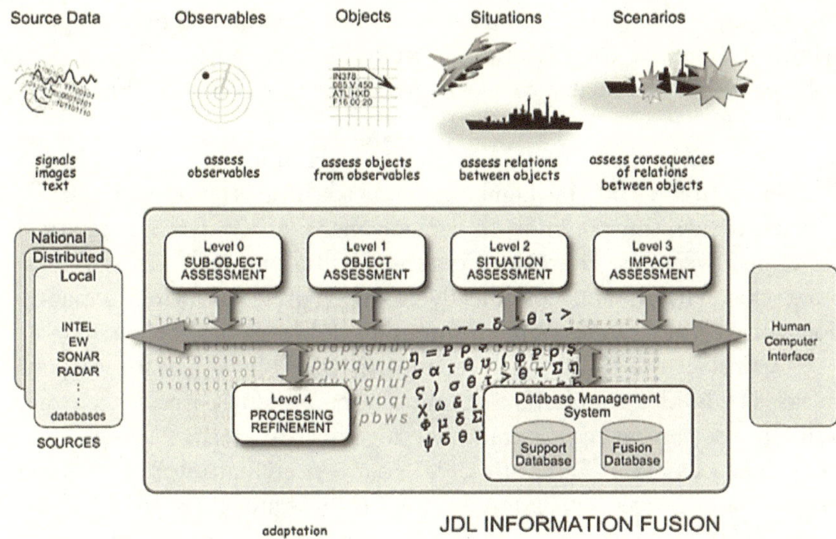

Figure 7.4. The Joint Directors of Laboratories Information Fusion Model.

tracking, typified by Kalman or particle filters in many military systems, or machine learning techniques for images, video, or speech signals. Figure 7.5 illustrates some state of the art, pixel registration performances

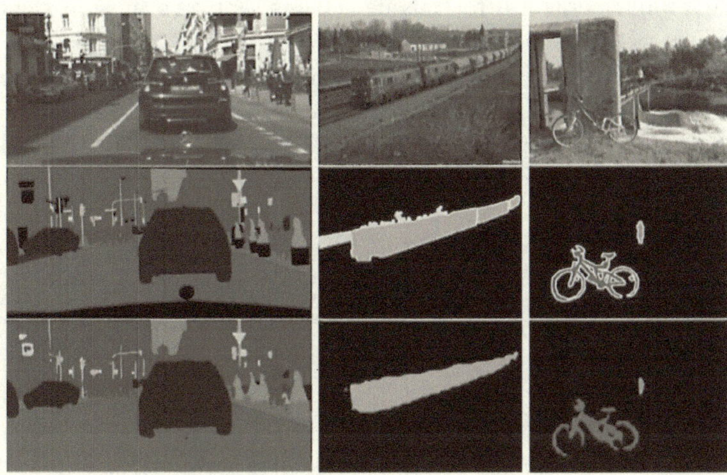

Figure 7.5. Examples of machine visual semantic segmentation from Wu et al. (2016) converted to gray scale. From top to bottom, each shows the original image, ground truth, and predicted labels.

on images, where the different colors indicate unique identified object names and include car, person, sign, train, bicycle, plants, pole/post, sky, and so forth. This relates to supporting ethical guideline 5, *Primacy for Human Life*.

A broad proposal to achieve higher-level fusion is proposed in the Consensus framework by Lambert, Saulwick, and Trentelman (2015). Consensus has the potential to provide many of the features required for ethical weapons. However, Consensus will also require the ability to learn epistemically and episodically to apply these hardwired semantic concepts to the environment in forming knowledge and beliefs.

Consider the benefit such technology would have had for weapons released by NATO warplanes that unintentionally struck a passenger train while it was passing across a railway bridge over the Južna Morava River at Grdelica Gorge in Serbia in 1999. Fourteen civilians were killed and another sixteen were wounded. Figure 7.6 illustrates three views of the scene as one weapon approached the strike location, which appears to be the bridge-supporting foundation. In future, semantic segmentation might pick up the presence of a train in sufficient time to permit the mission to be aborted or delayed.

However, that presence alone would not be enough for an ethical weapon. The machine would also need to process the meaning of this situation and its consequences. In the early stages of ethical weapon development, ethical weapons may have further mission-specific ethical packages implemented. If the missile is being sent on a mission to target a bridge, it would be loaded with the bridge ethical package installed. For a different mission it would use a different ethical package. So, although the missile would not have inbuilt ethics for all possible situations, it would have an inbuilt ethics package for the mission it is undertaking

Figure 7.6. Images as viewed through the weapon seeker video feed at three time points prior to impact.[9] In the left image the train can just be seen in the bottom left corner heading toward the bridge.

at the time, potentially making these individually oriented "morals." The protagonist then faces a choice between a dumb, undiscriminating missile; a smart missile with mission-specified ethics; or smart munitions at the hands of soldiers.

To bring our positive assertions into balance, a mission-specific package cannot necessarily deal with *every* conceivable situation. For example, a ruse of warfare used in Kosovo was the placement of ethnic Albanians as human shields under bridges and in the vicinity of military targets. Such actions were difficult for humans to identify and would likely also prove to be the case for ethical weapons. Indeed, considering the fact that the presence of innocent persons at military target locations need not be made obvious at all, with innocent people distributed randomly prior to an attack to confuse the attacker, makes it a deterrent once the consequences of an attack are discovered. However, we point out that the use of human shields is forbidden by Protocol I of the Geneva Conventions and is a specific-intent war crime as codified in the Rome Statute (ICC 2000). In the case of urban operations where noncombatants may be numerous and mixed among combatants, ethical weapons might be constrained only to prevent accidental strikes on protected objects and surrendering combatants due to the variability and complexity of unforeseeable situations.

To achieve machine learnable sentences with artificial neural network architectures, language sentences must be vector encoded. Hermann et al. (2015) describe a novel approach to developing real-world data sets that uses an attention mechanism to focus on critical aspects of the language. In ethical weapons that attention might be derived from "impact assessments" where any involvement of noncombatants is identified.

Machine learning approaches also need to be able to read and write to long-term memory for stored concepts and combine these seamlessly with inference. Memory networks (Weston et al. 2014) is one approach to this. Significantly richer inference processes will also be required. Ao et al. (2014) propose an inference process to automatically develop theories by abduction (derive hypotheses to explain observations), deduction (derive predictions from suggested hypotheses) and induction (generalize to new theories after testing the credibility/refutation of the hypotheses).

Ultimately, scalable and adaptable machine comprehension for ethical weapons may need to be achieved by *education* with symbolic concepts rather than training with data repetition, mirroring human acquisition of these capabilities through stories associating images and language in a pedagogical teacher–child relationship. This relates to ethical guideline 14, *Military Education and Training*.

1.1.3 Action

Cognitive routines in Attitude are scheduled through short-term memory, in order to affect the world through the weapons effectors. These effectors might include flight control surfaces, sensor directional steering, and the explosive payload (to neutralize or reduce the explosive yield).

Authority

We expect military commanders to retain authority over ethical weapons in all but certain special circumstances, but before we explain this exception, the nature of the future military enterprise must be considered along with some concepts from law. This relates to ethical guideline 8, *Military Commanders, Developers, and Defense Departments Are Accountable for Ethical Weapons*.

One question that arises is that of which commanders have authority over which ethical weapons and how is this to be managed? Ethical weapons might be viewed as a pooled resource with a plethora of capabilities available to authorized commanders to request dynamically. This may allow a more efficient model than, for example, the operator of a platform who as commander "owns" only those weapons carried. This supports operational weapon allocation and commitment prior to tactical use. None of this obviates the need for adherence to the law of armed conflict, which, of course, includes proportionality.

This integration between commanders and ethical weapons is proposed to be managed through an agreement process. In everyday transactions between people, agreements are a ubiquitous and intuitive notion. "Meeting of the minds" (*consensus ad idem*) is the term used in contract law to refer to the intentions to form a contract. Contract law brings a useful formalism to agreements as identified by Lambert and Lambert (2012), which also identified the need for cognitive agents to be given legal identity.

We further observe a recent proposal from the European Parliament (Delvaux 2016) that calls for robots to be classified as "electronic persons," in effect granting legal status to them as software entities that perform as agents in the (commercial/practical) legal sense of the term *agent*:

> Calls on the Commission, when carrying out an impact assessment of its future legislative instrument, to explore the implications of all possible legal solutions, such as: . . .
>
> f) creating a specific legal status for robots, so that at least the most sophisticated autonomous robots could be established as having the status of electronic persons with specific rights and obligations, including that of making good any damage they may cause, and applying electronic personality to cases where robots make smart autonomous decisions or otherwise interact with third parties independently.

Together these advances signal the potential for ethical weapons to employ a legal agreement protocol, which formally embeds the natural process of agreement within machines so that commanders may better control them, to "adhere to existing legal principles and while ensuring clarity in the contract formation process" (Lambert and Lambert 2012). The protocol covers stages of contract formation, contract performance, and contract remedies. The relevant semantic notions are at the social and intentional levels from figure 7.2 and include "agrees," "invites" (invitation to treat), "proposes," "offers," "accepts," and "rejects."

To aid in conceptualizing how such a system might operate, consider the similarity of these dynamic binding agreements to eBay:

> Command resembles the vendor expressing the intent of sale, with any member of the collective potentially being a vendor. Control resembles the process by which the purchaser acquires the sale item, with any member of the collective potentially being a purchaser. (Lambert and Scholz 2007)

Formally, and simplified from Lambert and Nowak (2008), an offer involves informing one's own intent for another to that other, while acceptance (signaling agreement) involves informing compliance to that intent to achieve outcome α:

$$\mathbf{offers}(Y,X,\alpha)=\mathbf{intends}(Y,\mathbf{intends}(X,\alpha))\&\mathbf{informs}(Y,X,\mathbf{intends}(Y,\mathbf{intends}(X,\alpha))$$

and

$$\mathbf{agrees}(Y,X,\alpha)=\mathbf{offers}(Y,X,\alpha)\&\mathbf{intends}(X,\alpha)\&\mathbf{informs}(X,Y,\mathbf{intends}(X,\alpha))$$

This characterizes X's agreement with Y to satisfy α through X's informed acceptance that X intends to satisfy α. An electronic legal agreement protocol provides all the necessary means to ensure legal accountability at every step, recording all communicative acts and states by the contracting parties, that is, between ethical weapons, between ethical weapons and commanders, and between commanders.

What does this binding of intent and action mean for authority in mixed initiative arrangements between commanders and their ethical weapons? From Lambert and Scholz (2007, 29), we assert that "an automated agent's *authority* is not determined by *a priori* rank, but depends upon the role it assumes in social agreements, given available competencies. In the end, authority is a matter of agreement." Thus, we conceive of ethical weapons representing a scope of outcomes that are limited by authorization. This scope may be narrow in some situations and broad in others, at the discretion and accountability of commanders.

To illustrate a broader scope example, several commanders may all have authority over several ethical weapons to strike enemy shipping within some limited time and space. The authority to strike enemy flagged ships means that in this example, ethical weapons are permitted to strike *any* enemy flagged ships within some limited time and space. However, ethical weapons would deny strike requests not within the time period, or beyond the space, or were there to be no enemy flagged vessels, or if a range of other violations were detected such as a frigate being alongside a civilian cargo ship—again, unless overridden by an authorized commander.

Resource allocation of ethical weapons may be managed at multiple levels of legal agency, to reduce the complexity of "contracting" otherwise expected from operating formal legal agreements at the level of the individual across the whole enterprise (Lambert and Scholz 2005). The legal sense of agency means that any agent who acts within the scope of authority conferred by their principal (another form of agent) binds the principal in the obligations they create against third parties. Agency is then defined in simplified form from Lambert and Nowak (2008), as agreements formed by Agent Z with third party Y to achieve outcome α on behalf of principal X are deemed to be agreements between X and Y to achieve outcome α:

agent_for(Z,X,Y,α)=agrees(X,Z,agrees(Z,Y,agrees(X,Y,α)))

Therefore, agency is in effect three contracts.

Figure 7.7 depicts management levels that may include the individual level where an officer (e.g., a soldier) manages their own ethical weapons allocation; the platform level where a platform commander (e.g., represented by a helicopter carrier) manages the collocated collective allocation on behalf of the ensemble of individuals; the team level where a team commander (e.g., represented by a soldier, a helicopter carrier, a jet fighter) manages their collective allocation on the basis of a shared common intent; and the coalition level (e.g., across government, international coalitions) where the coalition commander cooperates and competes for ethical weapon allocation with others but all are unified under common (agreed) principles.

Legal agency ensures authority and obligations propagate correctly throughout this enterprise. These four levels of management correspond to four kinds of principals. We expect ethical weapons to be able to assume the role of agent or third party but not the principal.

As an example, a principal at the coalition level may invoke agent team commanders to achieve an outcome. Providing they remain within the scope of the original outcome, they remain authorized to act. Some

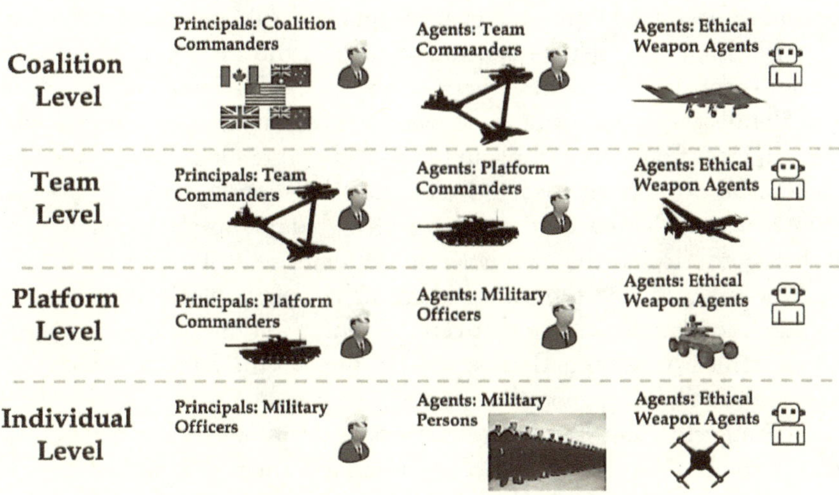

Figure 7.7. Management and agency of ethical weapons.

new outcomes still within the scope of the original outcome may in turn be generated by those teams, thus making them principals for other agent platform commanders to achieve, and so on. Of course, not all of these levels will be required in all situations.

Thus legal "agencies" may be created and dissolved dynamically at each level as agreements are attempted, created, and dissolved, with all contract states recorded for review of action to assure accountability across the entire enterprise.

This describes the creation of agency dynamically by the express appointment of a principal. If the agent exceeds their authority, then the principal will not be bound and the agent will be individually accountable to the third party for breach of authority. In contract law there are other forms of creation of agency than appointment by a principal. The "implied appointment by a principal," "apparent or ostensible authority," and "binding by estoppel" relate to informal cases of agency formation and might apply to agreements between commanders, but not to agreements between commanders and ethical weapons.

There is one other form of agency creation and that is termed "by necessity." This should apply to ethical weapons in the role of legal agent, when it exceeds its authority (scope of outcome) by intervening on behalf of the principal (a commander) in an emergency. The circumstances of necessity are important and may include the impracticality of the agent communicating with the principal (recalling ethical guideline 11, *Human on- and off-the-Loop*) if the emergency is time-critical, or if communication is not possible or is denied due to jamming of the weapon control link. Given these situations, the weapon agent should be treated as though it had the necessary authority to do what was reasonably necessary to save the principal from committing an illegal and unintended act. Specifically, the ethical weapon agent should have the express authority to *not achieve the outcome* requested by the commander, thereby operating outside of formal authority. In the case where an attempt to strike is aborted, but later goes ahead, the ethical weapon might eject a black box recorder for later communication or recovery to permit information on the state of the ethical weapon and target area related to the authority override, before the ethical weapon is destroyed.

Authority must be supported by adequate information security, and should not be undermined by a cyber attack. Information security technologies are required to establish authorized commanders of ethical weapons through authentication, supported by encryption technologies and secure key management. Should for any reason control over an ethical weapon be lost, a fail-safe mode of operation should limit the time for which the weapon may remain operationally active; the location(s) for weapon recovery; and no-go zones or geo-fencing. This functionality should be secured using low-complexity and high-reliability hardware separated from and managed independently of the rest of the weapon system to ensure it remains secure. This relates to ethical guideline 9 for the *Security of Ethical Weapons* and ethical guideline 13 for *Fail-Safe Management.*

Responsibility

We expect military commanders to retain responsibility over ethical weapons. However, as we propose ethical weapons to possess a kind of practically orientated legal agency, there will be an onus on these systems to explain their actions, with the owners of such weapons obligated to provide that explanation. Lambert and Scholz (2007) assert that "an automated agent's *responsibility* will follow from the social agreements it forms, given available competencies." Responsibility can be understood as commitment to an informed intention (Lambert and Nowak 2008). Individual X is responsible to individual Y to achieve outcome α if when X informs Y that it intends outcome α, which implies it actually intends it.

$$\mathbf{responsible} = (\mathbf{informs}(X,Y,\mathbf{intends}(X,\alpha) \Rightarrow \mathbf{intends}(X,\alpha))$$

The intent is bound into the legal agreement contract for all parties once acceptance is communicated to the offerer and a contract is formed as described earlier. Once formed, contract performance considers whether the legal obligations conferred at formation are met.

Ethical weapons will be responsible to commanders so that, for an outcome α:

$$\mathbf{responsible} = (\mathbf{informs}(\text{Weapon},\text{Commander},\mathbf{intends}(\text{Weapon},\alpha)) \Rightarrow \mathbf{intends}(\text{Weapon},\alpha))$$

Under normal conditions, the ethical weapon as agent fulfils its contract with the commander as principal and the agreement is satisfied. However, the essential test of responsibility is, in the event of an agreement being unsatisfied, how affected parties are to be remediated.

In the military, there is no duty to obey a command that is *manifestly* unlawful. Indeed, ethical autonomy would suggest there is a *duty not to obey*. This is the same inspiration for ethical weapons. The situation may occur as an intended war crime on behalf of the commander, or more likely due to the fog of war when the commander was not aware that the situation was different from what he imagined. Of course, the opposite case is also possible: for example, a Red Cross symbol on the side of a ship may be known to the commander as unlawful protective cover for enemy combatants. However, if this is known prior to ordering a strike, this awareness should be available to the ethical weapon to permit a lawful strike to continue. If it was not known to the ethical weapon, the default course of action would be to not carry out the strike. The number of cases in which uncertainty exists or whereby action is nullified might be expected to be reduced over time as protective symbols come to be accompanied by individualized, internationally (and securely) registered barcodes that would offer an enhanced level of safety and security in declared combatant and noncombatant military operations.

Of course, no technology is infallible. When ethical weapons are at fault, responsibilities in the military enterprise might extend to the need to incorporate a system to record the state of all ethical weapon legal agreements and relevant supporting information, which might later inform the payment of damages to affected persons and larger reparations payments to nations, as required under international law. This relates to achieving ethical guideline 10 for the *Awareness and Recording of Responsibility Transfers*. Sections 4.5 and 4.6 support ethical guideline 5, *Armed Conflict Shall Be Managed by Mixed-Initiative Agreements*.

There may also be a requirement for a technical system to provide explanations. It is worth noting that to date the most competent AI algorithms are also the least able to explain or justify their decisions, or both. This has led to the US eXplainable AI (XAI)[10] initiative led by the Defense Advanced Research Projects Agency. Considering the European Union's General Data Protection Regulation taking effect in 2018 and similar initiatives related to individual privacy, a "right to explanation" (Goodman and Flaxman 2016) may become the expected norm by the general public with a commensurate expectation to extend to military transparency about failures, for example identification of the fault(s), testing and assurance of proposed repairs, and adaptation of the fleet of software to repair affected ethical weapons.

Conclusion

No international ban on "weaponizing AI" will prevent terrorists, rogue states, state nonsignatories, and perhaps some state signatories from the development or use of autonomous weapons. However, activists might consider advocating ethical guidelines for weapons in order to minimize unintended weapon outcomes and institutionalize the means of weapons regulation, accountability, and remediation. As a consequence, this may reduce the motivation of relatives and friends of noncombatant victims to turn their energies into becoming future combatants.

The concept of ethical weapons as fully integrated elements of the military enterprise suggests that some decision-making qualities once considered unique to humanity be embedded within machines. This acknowledges that these systems remain part of a broader "system of control," in which the operational commander plays only a part (Government of Australia 2019). The commander could not reasonably be responsible or liable for a disaster if, for example, it is deemed to be due to a fault in weapon design, the data used to train the weapon, or improper certification acceptance. The intended effect is not to *cede* humanity so much as to *exceed* human-only decision-making through new socio-technical ecosystems with an even greater ethical capacity. We believe this is an imperative that the people of the world should expect and the militaries that serve them cannot ignore.

Notes

1. http://www.stopkillerrobots.org/.
2. http://www.stopkillerrobots.org/the-problem/.
3. One may argue that adversaries who know this might "game" the weapons by posing under the cover of "protection." If this is known, it is a case for (accountable) human override of the ethical weapon, and why we use the term "unexpected." Noting also that besides being an act of perfidy in the case of such use of protected symbols, which has other possible consequences for the perpetrators, it may in fact aid in targeting as these would be anomalies with respect to known Red Cross locations.
4. Baker and Pattison (2012) have argued analogously for the use of private military and security companies in humanitarian efforts.
5. http://www.stopkillerrobots.org/the-problem/.

6. This chapter focuses on ethical weapons broadly, which cover a spectrum from current munitions augmented with image recognition to prevent fratricide (which has been labeled "minimally just autonomy") to future weapon systems with more cognitive capabilities (Galliott and Scholz 2019). It is recognized that more complex weapon systems will require more complex ethical and epistemological frameworks. It is not the purpose of the present chapter to even sketch these, but it is worth nothing that it is an exciting area for future research.

7. Attitude is a cognitive modeling framework that makes use of propositional attitude expressions.

8. According to Lambert and Nowak (2008), "routine(R) identifies R as a (cognitive) routine, with use of the term 'routine' intending to appeal to both 'routine' in the sense of 'routine behaviour' and 'routine' in its Computer Science sense."

9. https://www.youtube.com/embed/t6zKEUGsPCo.

10. https://www.darpa.mil/program/explainable-artificial-intelligence.

References

Ao, Z., J. Scholz, and M. Oxenham. 2014. "A Scientific Inquiry Fusion Theory for High-Level Information Fusion." In *17th International Conference on Information Fusion (FUSION)*, edited by Juan M. Corchado, James Llinas, Jesus Garcia, Jose Manuel Molina, Javier Bajo, Stefano Coraluppi, David Hall, Moises Sudit, et al., 1–8. July. IEEE Explore. Berlin: Springer Verlag.

Arkin, R. 2009. *Governing Lethal Behavior in Autonomous Robots*. Boca Raton, FL: CRC Press.

Backstrom, A., and I. Henderson. 2012. "New Capabilities in Warfare: An Overview of Contemporary Technological Developments and the Associated Legal and Engineering Issues in Article 36 Weapons Reviews." *International Review of the Red Cross* 94(886): 483–514.

Barton, R. 2015. *The Weapons Detective: The Inside Story of Australia's Top Weapons Inspector*. Melbourne: Black Inc. Agenda.

Bolia, R. S., and R. E. Slyh. 2011. "Representation and Comprehension in Machine Translation and Intelligent Decision Support." *IEEE Intelligent Systems* 26(4): 40–47.

Bonnefon, J.-F., A. Shariff, and I. Rahwan. 2016. "The Social Dilemma of Autonomous Vehicles." *Science* 352(6293): 1573–76.

Bundesministerium für Verkehr und digitale Infrastruktur. 2017. "Ethik-Kommission automatisiertes und vernetztes Fahren." Berlin.

Copeland, D., and I. Reynoldson. 2017. "How to Avoid 'Summoning the Demon': The Legal Review of Weapons with Artificial Intelligence." *Pandora's Box* 24: 97–109.

Delvaux, M. 2016. *Draft Report with Recommendations to the Commission on Civil Law Rules on Robotics*. European Parliament Committee of Legal Affairs, 2015/2103(INL). Brussels.

Dennett, D. C. 1971. "Intentional Systems." In *Brainstorms: Philosophical Essays on Mind and Psychology*, edited by D. C. Dennett, 3–22. Brighton, Sussex, UK: Harvester Press.

Dennett, D. C. 2009. "Intentional Systems Theory." In *The Oxford Handbook of Philosophy of Mind*, edited by A. Beckermann, B. P. McLaughlin, and S. Walter. Oxford: Oxford University Press.

Doswald-Beck, L., ed. 1994. *San Remo Manual on International War Applicable to Armed Conflicts at Sea, 12 June 1994*. Prepared by international lawyers and naval experts convened by the International Institute of Humanitarian Law. Cambridge: Cambridge University Press. S 46(d). https://ihl-databases.icrc.org/applic/ihl/ihl.nsf/385ec082b509e76c41256739003e636d/7694fe2016f347e1c125641f002d49ce.

Endsley, M. R. 1988. "Design and Evaluation for Situation Awareness Enhancement." In *Proceedings of the Human Factors and Ergonomics Society Annual Meeting*, 97–101. Santa Monica, CA: Human Factors Society. https://journals.sagepub.com/doi/10.1177/154193128803200221.

Galliott, J. 2015. *Military Robots: Mapping the Moral Landscape*. London: Routledge.

Goddard, C. 2010. "The Natural Semantic Metalanguage Approach." In *The Oxford Handbook of Linguistic Analysis*, edited by Bernd Heiene and Heiko Narrog, 459–84. Oxford: Oxford University Press.

Goodman, B., and S. Flaxman. 2016. "European Union Regulations on Algorithmic Decision-Making and a 'Right to Explanation.'" arXiv preprint:1606.08813.

Government of Australia. 2019. "Australia's System of Control and Applications for Autonomous Weapon Systems." Paper submitted to the Group of Governmental Experts on Emerging Technologies in the Area of Lethal Autonomous Weapons Systems, Geneva, March 25–29. https://www.unog.ch/80256EDD006B8954/(httpAssets)/39A4B669B8AC2111C12583C1005F73CF/$file/CCW_GGE.1_2019_WP.2_final.pdf.

Hall, D. L., and J. Llinas. 1997. "An Introduction to Multisensor Data Fusion." *Proceedings of the IEEE* 85(1): 6–23.

Hermann, K. M., T. Kocisky, E. Grefenstette, L. Espeholt, W. Kay, M. Suleyman, and P. Blunsom. 2015. "Teaching Machines to Read and Comprehend." In *Advances in Neural Information Processing Systems*, 1693–1701. https://papers.nips.cc/paper/5945-teaching-machines-to-read-and-comprehend.pdf.

Human Rights Council. 2013. "Report of the Special Rapporteur on Extrajudicial, Summary or Arbitrary Executions, Christof Heyns." Twenty-third session, Agenda item 3, Promotion and Protection of All Human Rights, Civil, Political, Economic, Social and Cultural Rights, Including the Right to

Development. A/HRC/23/47. https://www.ohchr.org/Documents/HRBodies/HRCouncil/RegularSession/Session23/A-HRC-23-47_en.pdf.

International Committee of the Red Cross (ICRC). 2005. "Protocol Additional to the Geneva Conventions of 12 August 1949, and Relating to the Adoption of an Additional Distinctive Emblem (Protocol III)." December 8. https://ihl-databases.icrc.org/applic/ihl/ihl.nsf/INTRO/615.

International Criminal Court. 2000. "Elements of Crimes, Art. 8(2)(b)(xxiii)." U.N. Doc. PCINICC/2000/1/Add.2.

International Human Rights Clinic. 2012. "Losing Humanity: The Case against Killer Robots." Harvard Law School. https://www.hrw.org/sites/default/files/reports/arms1112ForUpload_0_0.pdf.

Jaidee, U., H. Muñoz-Avila, and D. W. Aha. 2011. "Integrated Learning for Goal-Driven Autonomy." *Proceedings of the Twenty-Second International Joint Conference on Artificial Intelligence—Volume Three.* IJCAI'11.

Kahneman, D. 2011. *Thinking, Fast and Slow.* New York: Farrar, Straus and Giroux.

Lambert, D. A. 1999. "Ubiquitous Command and Control." In *1999 Information, Decision and Control. Data and Information Fusion Symposium, Signal Processing and Communications Symposium and Decision and Control Symposium. Proceedings (Cat. No.99EX251)*, 35–40. Adelaide, South Australia. https://ieeexplore.ieee.org/document/754123.

Lambert, D. A. 2001. "Situations for Situation Awareness." Proceedings of the 4th International Conference on Information Fusion, Montreal, Canada.

Lambert, D. A. 2003. *Grand Challenges of Information Fusion.* Proceedings of the Sixth International Conference of Information Fusion, Cairns, Queensland, July 8–11, 213–19.

Lambert, D. A., and A. Lambert. 2012. "The Legal Agreement Protocol." In *High-Level Information Fusion Management and System Design*, edited by E. Blasch, É. Bossé, and D. A. Lambert, 173–90. Boston: Artech House.

Lambert, D., and C. Nowak. 2008. *The Mephisto Conceptual Framework.* Defence Science and Technology Organisation Technical Report, DSTO-TR-2162.

Lambert, D. A., and J. B. Scholz. 2005. *A Dialectic for Network Centric Warfare.* Proceedings of the 10th International Command and Control Research and Technology Symposium, MacLean, VA, June 13–16.

Lambert, D. A., and J. B. Scholz. 2007. "Ubiquitous Command and Control." *Intelligent Design Technologies* 1(3): 157–73. http://content.iospress.com/articles/intelligent-decision-technologies/idt00013.

Lambert, D. A., A. Saulwick, and K. Trentelman. 2015. "Consensus: A Comprehensive Solution to the Grand Challenges of Information Fusion." Proceedings of the *18th International Conference on Information Fusion*, 908–15. Washington, DC, July 6–9. https://ieeexplore.ieee.org/document/7266656.

McCann, C., and R. Pigeau. 1999. "Clarifying the Concepts of Control and of Command." Proceedings of the *Command and Control Research and Technology Symposium*, 475–90. Washington, DC: CCRP, Department of Defence.

Luetge, C. 2017. "The German Ethics Code for Automated and Connected Driving." *Philosophy and Technology* 30: 547–58.

Mikolov, T., K. Chen, G. Corrado, and J. Dean. 2013. "Efficient Estimation of Word Representations in Vector Space." arXiv preprint:1301.3781.

Miller, T., A. R. Pearce, and L. Sonenberg. 2017. "Social Planning for Trusted Autonomy." In *Foundations of Trusted Autonomy*, edited by Hussein A. Abbass, Jason Scholz, and Darryn J. Reid, 61–81. Studies in Systems, Decision and Control Series, vol. 117. New York: Springer International.

Nikitin, M. B., P. Kerr, and A. Feickert. 2013. Syria's Chemical Weapons: Issues for Congress, 30 September 2013. https://fas.org/sgp/crs/nuke/R42848.pdf.

Office of General Counsel, United States Department of Defense. 2016. *Department of Defense Law of War BICA 2010 Manual*. Washington, DC: Department of Defense.

Samsonovich, A. V. 2010. "Toward a Unified Catalog of Implemented Cognitive Architectures." In *BICA 2010*, edited by Alexei V. Samsonovich, Kamilla R. Johannsdottir, Antonio Chella, and Ben Goertzel, 195–244. Amsterdam: IOS Press.

Scheutz, M. 2017. "The Case for Explicit Ethical Agents." *AI Magazine* 38(4).

Scholz, J., D. Lambert, D. Gossink, and G. Smith. 2012. "A Blueprint for Command and Control: Automation and Interface." In *Information Fusion (FUSION), 2012 15th International Conference on Information Fusion*, 211–17. July, IEEE.

Sparrow, R. 2016. "Robots and Respect: Assessing the Case against Autonomous Weapon Systems." *Ethics & International Affairs* 30(1): 93–116.

Teney, D., and A. V. D. Hengel. 2016. "Zero-Shot Visual Question Answering." arXiv preprint:1611.05546.

Tuffley, D. 2017. "At Last: The World's First Ethical Guidelines for Driverless Cars." *The Conversation*, September 3. https://theconversation.com/at-last-the-worlds-first-ethical-guidelines-for-driverless-cars-83227.

United Nations General Assembly, Human Rights Council. 2013. "Report of the Special Rapporteur on Extrajudicial, Summary or Arbitrary Executions, Christof Heyns." A/HRC/23/47, April 17.

Vanderelst, D., and A. Winfield. 2018. "An Architecture for Ethical Robots Inspired by the Simulation Theory of Cognition." *Cognitive Systems Research* 48: 56–66.

Vattam, S., M. Klenk, M. Molineaux, and D. W. Aha. 2013. *Breadth of Approaches to Goal Reasoning: A Research Survey*. Washington, DC: Naval Research Lab.

Weston, J., S. Chopra, and A. Bordes. 2014. "Memory Networks." arXiv preprint:1410.3916.

White, A., and M. Paterson. 2017. "Nuke Kid in Town: How Much Does the Treaty on the Prohibition of Nuclear Weapons Actually Change?" *Pandora's Box* 24: 141–56.

Wu, Z., C. Shen, and A. van den Hengel. 2016. *Wider or Deeper: Revisiting the ResNet Model for Visual Recognition*. November. arXiv preprint:1611.10080.

Conclusion

The Future (Idea) of Just War and Autonomous Weapons Systems

Steven C. Roach

As autonomous weapons systems become more widespread, they will continue to raise difficult moral questions about the role of human intervention in modern warfare. Today there are plans for using quantum sensors to detect moving targets and to attack and destroy satellites. These and other AWS are part of a changing twenty-first-century arsenal that will define the new landscape of future warfare. At the same time, they offer new opportunities for rethinking the morality of warfare. As Christian Brose (2019, 133) puts it, "autonomous systems will enable humans to spend less time on mental problems and more time on moral ones." Future high-tech AWS will not simply be more targeted and surgical: they will also shift the burden from humans to AWS in terms of reducing pain and suffering. This shift requires us to consider more seriously the ethical content of AWS through artificial intelligence (AI) testing and ethical theorizing that supports and extends just war principles. In short, we will need to formulate moral concepts that help us to understand the disparities and connections between just war theorizing and the empirical knowledge of AI.

As one of those concepts, dual moral responsibility allowed us to rethink some of the ethical gaps between just war theory and AWS, particularly the refusal to extend ethics to robots. As David Gunkel showed in his chapter, the refusal to consider the moral conduct of robots represents the bad faith (denying the premises of truth) in the potential of robot agency. Rather than devising an ethics of robots, philosophers and policymakers have affirmed an unquestioned belief in the exclusivity of human morality, dismissing the possibilities of moral autonomy in robots. What this points to is the status quo scenario in which denying the rights of robots will become an increasing moral liability, especially given that the adaptive capacity of robots is a seeming building block of knowledge and, by extension, consciousness.

A more positive scenario, by contrast, suspends this anthropocentric bias and encourages engagement with the possibilities of robots' moral and legal agency. It assumes that the conventional limits of discerning robots' moral intelligence are neither fixed nor stable. Rather, such intelligence may be evolving and may one day rival human intelligence. As Peter Singer (2009, 403) writes, "Robots have great difficulty interpreting context, and, at least, until they match humans in intelligence, it simply does not make sense to interpret a machine as having the equivalent of human rights or self-defense." Humans may not be machines per se, but to think that there can be no equivalent to human intelligence is to underestimate the adaptive nature of AI. If science fiction is any indication, we tend to downplay how fiction shapes the reality of warfare. And yet science fiction is what allows us to imagine a different world. But even it cannot escape its own anthropocentric bias against robots' intelligence.

Consider for example the theme of world domination by machines (i.e., cyber-borgs that are more human than today's robots), which is depicted in films, such as the *Terminator* series, and in literature as far back as Isaac Asimov's *I, Robot*, first published in 1950. When robots take over the world, they pose an existential threat to humanity through their cold calculation and lack of emotion. As some of the contributors to this volume have suggested, this threat does not lie in the machines per se, but rather in the humans who created them. That is, the robots' menace is a projection of the threat that lies deep within the human psyche; it must always find an escape or sublimating object. Moreover, such projection does not diminish with time: it continues to intensify with human technological ingenuity and with fears and anxieties about controlling the future. Fear, in short, predisposes us to think of AWS as objects of displaced pleasure and pain, rather than as subjects.

Still, science fiction can also tell a story in which machines are treated as alternative subjects possessing rudimentary, albeit fractured notions of responsibility and self-awareness, such as Steven Spielberg's *A.I., Artificial Intelligence* (2001). The film depicts competing notions of moral responsibility between the self-aware machines that struggle against their fated destruction by humans, and humans who have to dispose of (the) malfunctioning robots. It's a tension that suggests an inversion of the cold, calculating machines bent on the domination of humans. In the end, when we champion the moral authority of humans, we also dismiss this tension and reinforce a narrow moral space in the imagination (Roach 2019).

The point, then, is that we simply do not know enough about this moral space to justify the fixed limits of nonhuman intelligence (in relation to human intelligence). It makes more sense to imagine an expanding moral space, that is, an open-ended notion of a common good for which humans and nonhumans assume complementary moral roles in preserving peace and justice. In much the same way that humans need to further accommodate nonhuman animals' rights (i.e., recognizing a full schedule of animal rights), so too do we need to imagine a world in which robots are imagined as playing a positive moral role.

Which brings us to the time and evolution underlying the notion of dual moral responsibility. So far, I have suggested that rapid advances of technology will outpace our ability to understand and explain the moral agency of robots, and that this in turn will require an equally quick transition to a shared moral responsibility among humans and nonhumans. However, moral thinking tends to operate at a relatively slow pace in terms of achieving efficacy. Robert Sparrow, for instance, sees this process as a crucial test of looking outside the conventional moral box or accepted rules and principles of just war theory in relation to AWS. It takes time, in other words, to move beyond our long-standing anthropocentric bias against nonhuman moral agency and to consider more seriously the moral stakes of excluding AWS from just war theorizing (i.e., that moral wars or conflicts will be fought by entities whose autonomy lies outside the scope of the rules of war). Yet if we fail to address these rising stakes, we also risk assigning just war theory to the heap of history, or rendering it less relevant and effective in the future. As Sparrow (2015, 109) puts it:

> I have suggested that widespread public revulsion at the idea of autonomous weapons should be interpreted as conveying the belief that the use of AWS is incompatible with respect. If I

> am correct in this, then even if an interpersonal relationship may be held to exist between the commanding officer who orders the launch of an autonomous weapon system and the individuals killed by that system, it should be characterized as one of disrespect.

As this passage suggests, broadening respect is one way of gauging the changing dynamics of war in terms of carrying out the rules of moral conduct in war, including the decisions that affect humans' ability to act in accordance with such rules. If anything, it means entertaining the prospect that AWS will achieve some limited capacity for making moral decisions in war, such as the capability to distinguish among "soft targets" and to uphold the principle of proportionality.

It would seem, then, that the ethicists most closely engaged with the practical applications of robot intelligence would be among the most cautious or skeptical of robot responsibility. But as Jason Scholtz, Robert Bolia, Dale Lambert, and Jai Galliott have shown in their chapter, such ethicists remain both optimistic and open-minded about the ethical future of AWS. In their research, they speculate and hypothesize about the virtues and moral possibilities of AWS and demonstrate how practical knowledge of AWS should dispel the fear of LAWS and encourage more grounded ethical thinking of AWS. Such research is important for two reasons. First, it illustrates some of the key intersecting points of just war ethics and AWS (or ethical procedures for AWS), which provide implications for merging the ethical tracks mentioned above. Second, it again exposes the anthropocentrism underlying revisionist war theory, to the extent that human rationality and emotion are part of the adaptive capacity of robotic intelligence that might help justify rights and a uniform moral code of law. In short, human rationality and intelligence are not diametrically opposed to one another. Rather, rationality is wedded to emotion through the adaptive behavior that drives all living species consciousness. If such collective consciousness reflects the adaptive capacity of human and nonhuman animals, then AWS' relatively rudimentary adaptive capacity should be considered as a building block of moral agency or at least a condition of thinking about this agency.

As Gunkel states, "a robot's rights are not unthinkable, but their consideration is tightly constrained by and limited to the rights that belong to its human use" (Gunkel 2018, 48). Moral dogmatism (in the area of the moral study of robots) essentially strands our imagination in a logic in which only humans can simultaneously violate the laws designed (by them) to protect them from such violence, and only humans can be the victims of violence.

This leaves us with the important question of how ceding control to robots may well be a good thing, particularly if it leads to a more sweeping and uniform notion of moral conduct in future wars. But the full answer will depend on a gradualist approach to the practical ethics of AWS. As Jai Galliott argued in his chapter, AWS or robots, or both, function independently in ways that reaffirm a growing adaptive capacity with their own limits, the building block of experience.

Finally, the idea of integral responsibility represents a stage beyond dual moral responsibility—where the rights of nonhumans mutually complement one another to reduce the violent effects of warfare. As such, it constitutes a heterodox vision or configuration of a future ethics of twenty-first-century warfare and requires ethics scholars to contest the anthropocentric bias against AI: in short, to work toward an encapsulating ethics that merges the evolving dual tracks of human and nonhuman responsibility and that engages the possibility of shared moral responsibility in warfare.

References

Brose, Christian. 2019. "The New Revolution in Military Affairs." *Foreign Affairs* 98(3): 122–34.

Gunkel, David. 2018. *Robots' Rights*. Cambridge, MA: MIT Press.

Roach, Steven C. 2019. *Decency and Difference: Humanity and the Global Challenge of Identity Politics*. Ann Arbor: University of Michigan Press.

Singer, Peter W. 2009. *Wired for War: The Robotics Revolution and Conflict in the 21st Century*. London: Penguin.

Sparrow, Robert. 2015. "Robots and Respect: Assessing the Case against Autonomous Weapons Systems." *Ethics and International Affairs* 30(1): 93–116.

Contributors

Robert Bolia received a BA in Mathematics from Wright State University and an MA in Military Studies (Joint Warfare) from American Military University. He is also a graduate of the US Air War College and the US Naval War College. From 1989 to 2008 he was a Research Scientist with the Air Force Research Laboratory, Human Effectiveness Directorate, USA. From 2008 to 2016 he served as an Associate Director of the Office of Naval Research Global in Tokyo, Japan, and Santiago, Chile. In 2016 he became Group Leader, Human Factors, in the Aerospace Division at the Defence Science and Technology (DST) Group, Australia. Since 2018 he has been Research Leader, Aerospace Systems Effectiveness at DST. He also leads the DST Strategic Research Initiative on Trusted Autonomous Systems.

Thomas E. Doyle II is an Associate Professor of Political Science at Texas State University. His research focus encompasses nuclear ethics, security studies, and international political theory. He has published in journals such as *Ethics and International Affairs*, *Global Governance*, *International Theory*, *Strategic Studies Quarterly*, *Ethics and Global Politics*, *Journal of International Political Theory*, *and Critical Military Studies*. He has also published two books: *Nuclear Ethics in the 21st Century: Survival, Order, and Justice* (2020) and *The Ethics of Nuclear Weapons Dissemination: Moral Dilemmas of Aspiration, Avoidance, and Prevention* (2015).

Amy E. Eckert is Professor of Political Science at the Metropolitan State University in Denver. Her book *Outsourcing War: The Just War Tradition in the Age of Military Privatization* (2016) addresses the many ethical challenges of ceding control to private military companies. She was also the section leader of the International Studies Association's International Ethics section.

Jai Galliott leads the Values in Defence and Security Technology Group at the University of New South Wales at the Australian Defence Force Academy. He also holds appointments at the Modern War Institute at the United States Military Academy, West Point, and the Centre for Technology and Global Affairs at the University of Oxford. He is an expert on the ethical, legal, and social implications of emerging military technologies including autonomous weapons and cyber systems. He is the author of *Military Robots: Mapping the Moral Landscape* (2015) and *Force Short of War in Modern Conflict* (2019).

David J. Gunkel specializes in the philosophy of technology. He is the author of over seventy-five scholarly articles and ten books, including *Thinking Otherwise: Philosophy, Communication, Technology* (2007), *The Machine Question: Critical Perspectives on AI, Robots, and Ethics* (2012), and *Robot Rights* (2018). He currently holds the position of Distinguished Teaching Professor of Communication Technology at Northern Illinois University. For more information, see http://gunkelweb.com.

Dale A. Lambert is the Executive Chair of the Technical Cooperation Program Command, Control Communications and Information Group, a long-standing five-nation cooperative effort on defence science and technology. He received a Bachelor of Science in Computer Science, a Bachelor of Arts in Philosophy, a Bachelor of Arts in Mathematics, a PhD in Artificial Intelligence, and Graduate Certificate in Management. He is the Chief of the Cyber and Electronic Warfare Division at the Defence Science and Technology Group, Australia. He specializes in higher-level data fusion research and has designed and implemented an Artificial Intelligence system for Sweden's air defence that was on sold to several countries.

Sommer Mitchell is an Assistant Teaching Professor of Global Studies and the Undergraduate Director of Global and International Studies at Pennsylvania State University. Her research focuses on international security, legitimacy, and private military and security companies. Sommer received her PhD in Government from the University of South Florida. She has more than ten years' teaching experience and has taught courses on international relations, globalization, international human rights, and global conflict and has extensive experience in faculty development through her work at USF's Global Citizens Project.

Steven C. Roach is Professor of International Relations and Graduate Director in the School of Interdisciplinary Global Studies at the University of South Florida. Among his recent books are *Handbook of Critical International Relations* (ed.) (2020), *Decency and Difference: Humanity and the Global Challenge of Identity Politics* (2019), *The Challenge of Governance in South Sudan: Corruption, Peacebuilding, and Foreign Intervention* (eds.) (2019), *Critical Theory of International Politics* (2010), and *Governance, Order, and the International Criminal Court: Between Realpolitik and a Cosmopolitan Court* (ed.) (2009).

Jason Scholz received a PhD in electrical engineering from the University of Adelaide and a degree in electronic engineering from the University of South Australia. He has over seventy-five refereed open publications and patents, in telecommunications, signal processing, artificial intelligence, and human decision-making. He is Professor (adjunct) at the University of New South Wales. He leads research, development, and showcasing of high-impact technologies in Trusted Autonomous Systems for persistent autonomy, machine cognition, and human–machine integration in close partnership with overseas governments, academia, and industry.

Laura Sjoberg is Professor of Political Science at the University of Florida and British Academy Global Professor of Politics and International Relations at Royal Holloway, University of London. Her research on gender and war has been published in more than three dozen journals in political science, gender studies, geography, and law. Sjoberg is author or editor of more than a dozen books, including, recently, *International Relations' Last Synthesis* (with J. Samuel Barkin, 2019) and *Handbook on Gender and Security* (with Caron E. Gentry and Laura J. Shepherd, 2018).

Peter Sutch is Professor of Political and International Theory at Cardiff University and a visiting Professor at the University of the Witwatersrand in South Africa. He writes on international political thought and on questions of just war, international law, and global justice, and is the author of *Ethics, Justice and International Relations: Constructing an International Community* (2001).

Index

www.ingramcontent.com/pod-product-compliance
Lightning Source LLC
LaVergne TN
LVHW050151080826
844660LV00002B/162